TRUTH, RATIONALITY, COGNITION, AND MUSIC

PHILOSOPHICAL STUDIES SERIES

VOLUME 102

The titles published in this series are listed at the end of this volume.

TRUTH, RATIONALITY, COGNITION, AND MUSIC

Proceedings of the Seventh International Colloquium on Cognitive Science

Edited by

KEPA KORTA

and

JESÚS M. LARRAZABAL

Institute for Logic, Cognition, Language and Information, The University of the Basque Country, Spain

KLUWER ACADEMIC PUBLISHERS

DORDRECHT / BOSTON / LONDON

A C.I.P. Catalogue record for this book is available from the Library of Congress.

ISBN 1-4020-1912-2

Published by Kluwer Academic Publishers,
P.O. Box 17, 3300 AA Dordrecht, The Netherlands.

Sold and distributed in North, Central and South America
by Kluwer Academic Publishers,
101 Philip Drive, Norwell, MA 02061, U.S.A.

In all other countries, sold and distributed
by Kluwer Academic Publishers,
P.O. Box 322, 3300 AH Dordrecht, The Netherlands.

Printed on acid-free paper

Printed in the Netherlands.

Contents

Preface

> A speech for the defence in a Paris murder trial, a road-safety slogan, Hobbes' political theory; each appeals to reason *of a kind*, but it remains an oblique and rhetorical kind. Each relies on comparisons rather than on direct statements, and none can override or supersede the conclusions of ethical reasoning proper. Nevertheless, just as slogans may do more for road safety than the mere recital of accident statistics, or of the evidence given at coroners' inquests, so the arguments of a Hobbes or a Bentham may be of greater practical effect than the assertion of genuinely ethical or political statements, however true and relevant these may be.
>
> Stephen Toulmin, *Reason in Ethics*, 1950.

The International Colloquium on Cognitive Science (ICCS), held in Donostia – San Sebastián every two years since 1989, has become one of the most important plazas for cognitive scientists in Europe to present the results of their research and to exchange ideas. The seventh edition, co-organized as usual by the Institute for Logic, Cognition, Language, and Information (ILCLI) and the Department of Logic and Philosophy of Science, both from the University of the Basque Country, took place from May 9 to 12, 2001, addressing the following main topics:

1. Truth: Epistemology and Logic.
2. Rationality in a Social Setting.
3. Music, Language, and Cognition.

4. The Order of Discourse: Logic, Pragmatics, and Rhetoric.

Around one hundred people from all over the world participated in ICCS-01. This volume contains the invited papers and tutorials presented there, as well as Martin Kusch's contributed paper, "Meaning Finistism and Truth," that merited the IBERDROLA Best Paper Award, given in 2001 for the third time. Let us briefly describe the papers one by one.

The first chapter, John Biro's "Cognitive Science and David Hume's Science of the Mind," is the only chapter in this volume devoted to the history of cognitive science, in commemoration of the 225th anniversary of the death of the great Scottish philosopher. Contemporary cognitive science is coupled with Hume's theory of mind. But, among all the figures of the history of philosophy that could be susceptible of study for the history of cognitive science, why does he choose Hume (besides the commemorative reason)? Was not he "academically" sceptic about any possibility of knowledge? So, why should he be relevant to cognitive scientists today? Biro argues that, far for being sceptical about the possibility of a science of the mind, Hume was maybe the first to consciously undertake the project of constructing one. Moreover, Hume anticipated some of the goals and some of the results of contemporary cognitive scientists. According to Biro, we should become aware of the fact that "it is an irony that the cognitive science of our day is sometimes explicitly contrasted with Hume's science of man, rather than being recognized as the latter's descendant," and that Hume is "the philosophical father of cognitive science".

The second chapter, Scott Soames' "Truth and Meaning: The Role of Truth in the Semantics of Propositional Attitude Ascriptions," focuses on the shortcomings that some semantic theories present when addressing the meaning of propositional attitude reports. The semantic theories taken under consideration are contemporary truth-conditional theories. According to Soames, these theories that try to specify the meanings of propositional attitude ascriptions, as they stand, are doomed to failure. However, he finds a way to extend these Davidsonian theories making them able to provide some information about the meanings of propositional attitude ascriptions. Nonetheless, he also finds that there are serious difficulties in extracting substantive testable consequences from even these extended theories.

The third chapter, Timothy Williamson's "Truth and Borderline Cases," adresses the relation between the derivability of the principle of bivalence, challenged by borderline cases for vague expressions, and some principles about truth and falsity. Starting with the latter, he replaces the usual disquotational characterization of truth by one that makes room for contextual variation in what is said by a sentence. He then explores the implications this characterization has for bivalence. Bivalence requires a principle of uniformity in what a given sentence says in a given context, and even though borderline cases for vague expressions may appear to provide reasons to deny such a uniformity, Williamson argues that it is not so: "The apparent loophole in the classical case for bivalence is ... merely apparent", he concludes.

The fourth chapter, Martin Kusch's "Meaning Finitism and Truth," reconstructs and defends a theory of linguistic meaning developed since the late seventies by Barry Barnes and David Bloor that has been called "Meaning Finitism", exploring at the same time some of its consequences for our understanding of truth. Meaning finitism views meanings as continuously made and re-made by language users. Like Barnes and Bloor, Kusch combines meaning finitism with meaning collectivism, and, perhaps more explicitly than the formers, he defends one of the most important philosophical consequences of that "strong" position: relativism concerning truth. This paper, which closes the bunch of papers connected with truth, got the IBERDROLA Best Paper Award at ICCS-01.

Manuel Liz's "Subjective Experience and the External World: Gaps and Bridges" is the first of the three papers in the volume, chapter 5 to 7, devoted to the philosophy of cognitive science. He attempts to bridge what he calls the ontological and the epistemological gaps between our subjective experiences and the external world. The epistemological gap arises from the problem of knowing the external world given that "it always seems possible to conceive our subjective experience as remaining the same through many kinds of changes in the external world"; the ontological gap arises when trying to understand how the external world can produce subjective experiences given the fact that "it seems always possible to conceive the external world —especially our brains and bodies— as remaining the same through many kinds of changes in our subjective experience." According to Liz, taking seriously the ontological gap entails being able to bridge the epistemological one and the strategy he proposes to bridge the latter could also be successfully applied to the former.

Chapter 6, Marcelo Sabatés' "The Explanatory Relevance of Psychological Properties," studies epiphenomenalist approaches to mental properties and explores their consequences for the explanatoriness of such properties, and "perhaps more generally," tells Sabatés, "of other special science properties." According to many critics, epiphenomenalist approaches to mental properties end up concluding that they are explanatory irrelevant or even that they are not real properties. Sabatés attempts to show that the indirect and direct arguments going from epiphenomenalism to lack of explanatory relevance, though compelling, are not conclusive, and he certainly shows which should be the main concerns on the issue. As he claims, this is a necessary first step "towards a theory allowing us to *realistically* explain with *real* epiphenomena."

Chapter 7, "Epistemology and Cognitive Theorizing" by Jesús Ezquerro and Fernando Martínez-Manrique, treats the mutual influence of studies in cognitive science, on the one hand, and theoretical projects in epistemology. The paper is divided into two parts. The first part traces a path through normative issues in the philosophy and history of science that grounds the view that the only plausible understanding of epistemology comes from a naturalist position. The second part of the paper studies the impact that a putative "folk epistemology" capacity (i.e., the natural ability to evaluate epistemic situations) could have on epistemology. This paper closes the series of papers on the philosophy of cognitive science.

Chapters 8 and 9 are concerned with the relation of music to cognition. The question made by Peter Kivy in the title of his paper, "Music, Language and Cognition: Which One Doesn't Belong," is clearly answered by the author: *language* doesn't belong. Music is not language. The first part of the paper explores the analogy between music and language. A wrong understanding of this analogy has caused pretty much confussion. Kivy insists that, though language-like in certain respects, music "is not language: it is not a language or part of a language." On the other hand, he claims that music and cognition belong together. More specifically, in the sencond part of the paper, he argues that the enjoyment and appreciation of music are concept-laden (but content-less) activities and, thus, deeply involved with cognitive processes.

While Kivy leaves it explicitly aside, Lairs Addis is primarily concerned with unconscious (or innate) knowledge in relation with music in his "Music and Knowledge." He adresses two questions that seem to involve some kind of unconscious knowledge. The first concerns knowledge "that we *bring to* music," what Addis calls "knowledge-*for*" listening to music; the second concerns knowledge that we, humans, get from music ("knowledge-*from*" listening to music). These two notions are also analysed in the light of a better-known philosophical distinction among knowledge-*that*, knowledge-*how*, and knowledge-*of*, as well as attending to music psychology with the aim of clarifying why music has the importance it has for human beings.

Chapter 10, "Event Coreference and Discourse Relations" by Laurence Danlos and Bertrand Gaiffe, is a paper on discourse semantics within the framework of the Segmented Discourse Representation Theory (SDRT). In particular, they focus on event coreference, which is shown to be the keystone for the correct analysis of four types of dicourses: particularizing, generalizing discourses, on the one hand, and explanation and result discourses, on the other. Their analysis allow them to present and explain some unusual linguistic phenomena (such as coreference between existentially quantified elements, or asymmetrical behavior of explanation and result discourses), as well as to shed new light on the discourse relations involved.

The last two chapters address the issue of rationality in a social setting. Chapter 11, "Rationality in Context" by J. Francisco Morales, approaches the problem from the perspective of social psychology. To begin with, he discusses the very concept of rationality for undermining the alleged irrationality in some cases of social behavior. In particular, he is concerned with the possibility of reconciling rational choice theory with the influence of culture in human decision-making. With this purpose, he discusses a couple of cross-cultural studies on conflict management strategies and on the obedience to authorities, respectively. The author concludes that "paying attention to [the context] is a necessary step to move beyond the rational-irrational dichotomy to a more ... adequate conceptualization of human behavior."

Chapter 12, "Individual and Collective Rationality in a Social Framework" by Luis A. Pérez Miranda, presents and explores the very roots of rational choice theory and game theory. The first part of the paper is devoted to the discussion of the main problems of individual decision-making from the viewpoint of instrumental rationality. The second part examines the efforts to extend rational choice theory when there is more than one individual in the decision-making, that is, when we have a social or even a collective decision to be made. Is game theory suitable for explaining phenomena like coordination, cooperation, or the emergence of norms? Pérez Miranda argues that it needs a deep revision that, on the one hand, takes into account the role played by structural rationality and, on the other hand, gets "less idealised models sensitive to experimental data, or ... models that find their scientific support in accurate computer simulation. "

As usual, this collection of papers selected from the ICCS-01 reflects, not only the diversity of the interdisciplinary origins and standpoints of the researchers who attended the Colloquium, but also serves as a proof of the richness, fruitfulness, and diversity of research in Cognitive Science today.

Kepa Korta and Jesus M. Larrazabal

Chapter 1

COGNITIVE SCIENCE AND DAVID HUME'S SCIENCE OF THE MIND

John Biro
University of Florida

What reason is there for coupling one of the newest branches of modern science with a long-dead philosopher, however august? And why with Hume, of all people? Was he not a sceptic, famous for questioning the possibility of any, and, thus, *a fortiori*, scientific, knowledge? Interesting as his reasons for such doubts may be to other philosophers, can they not be safely ignored by scientists, those actually engaged in pursuing and, surely, gaining, the knowledge he said was unattainable? But, then, what else has he to say to the latter? In what follows, I shall argue that far from being sceptical about the possibility of a science of the mind, Hume was perhaps the first to embark self-consciously on the project of constructing one. Furthermore, some of the goals and some of the results of his project anticipate in interesting and rarely noticed ways those of recent cognitive science and its philosophy.

1.

For almost two centuries after its appearance, Hume's philosophy was construed as essentially, perhaps entirely, negative.[1] Its goal was seen as

[1] With the possible exception of some of the practical philosophy. But even where these were read at all, the emphasis was typically on the sceptical or destructive passages (often taken

K. Korta and J. M. Larrazabal (eds.), Truth, Rationality, Cognition, and Music, 1-20.

showing that the unavoidable outcome of following through on some seemingly compelling empiricist principles was complete scepticism. That Hume should be read in this way over a stretch of time during which epistemological questions in general, and the sceptical threat in particular, dominated philosophy is hardly surprising. It is a common-place that with Descartes these questions had moved to center stage in a way that would have seemed, until then, quite unnatural to most philosophers since the ancient sceptics. Thus it is only to be expected that Hume should have been seen as engaged in the general post-Cartesian enterprize of assessing—in his case, apparently negatively—the prospects for human knowledge. It is also true that the first book of *A Treatise of Human Nature*, and *An Enquiry into Human Understanding*, the two parts of his *ôuvre* on which most interpreters have focussed until recently, contain a barrage of arguments that certainly seem to have been designed with a sceptical intent.

Thus Hume came to be seen as the arch-sceptic of the modern age. That such extreme scepticism is straightforwardly incompatible with the project Hume explicitly announced in the introduction to the very work which was interpreted as its expression was either not noticed or dismissed by the simple expedient of not taking him at his word.

In the Introduction to the *Treatise*, in what is one of the best-known passages in all his writings, Hume outlines a project it is hard to see a sceptic taking seriously: that of establishing a new science. He calls it "the science of man," and describes it as an investigation into "the nature of the ideas we employ, and of the operations we perform in our reasonings." Such an investigation is needed if any progress is to be made in any of the other sciences, since "all the sciences have a relation ... to human nature," and "lie under the cognizance of men, and are judged of by their powers and faculties." Thus, understanding the human mind is the key to understanding anything else:[2]

> There is no question of importance whose decision is not compriz'd in the science of man; and there is none, which can be decided with any certainty, before we become acquainted with that science. In pretending therefore to explain the principles of human nature, we in effect propose a compleat system of the sciences, built on a foundation almost entirely new, and the only one upon which they can stand with any security. [T xx]

---

out of context), such as the notorious one about the impossibility of deriving an 'ought' from and 'is'. (*A Treatise of Human Nature*, pp. 469-70. References to this work in what follows will take the form 'T pp.')

[2] For a more thorough discussion of this claim than is possible here, see Biro (1993). See also Monteiro.

Explaining the principles of human nature involves "examining the Mind... to discover its most secret Springs and Principles." Though these may "lie very deep and abstruse", the new "experimental method of Reasoning", modelled after that used with such spectacular success by Boyle, Newton and others in what might be called the new science of matter, holds out the hope of results no less far-reaching in the domain of "moral subjects".[3] The "total alteration" and "revolution" Hume says this new science brings to the intellectual scene consists in becoming an "anatomist" of human nature. The secret principles that "anatomizing human nature in a regular manner" enables us to discover need not, however, be *occult*; they are accessible to an investigator using the right method, a method that contrasts with that of the metaphysician. The latter's is mere arm-chair speculation, yielding "many chimerical systems" in which "hypotheses [are] embrac'd merely for being specious and agreeable," rather than because they can be shown to be true, or even probably true, by some objective method.

But what method? As an empiricist, Hume can allow no source of knowledge other than experience and recognizes no claim to knowledge not based on experience: "The only solid foundation we can give to this science... must be laid on experience and observation." (T xx). No science "can go beyond experience, or establish any principles which are not founded on that authority." Thus the right method calls for "exact experiments," aiming at "render[ing] all our principles as universal as possible". This, in turn, requires "tracing up our experiments to the utmost, and explaining all effects from the simplest and fewest causes." (ibid.)

The kind of speculative metaphysics with which Hume contrasts his new science poses questions to which no principles derived from experience can yield an answer. It pretends to offer deeper and more general principles based on pure reason and claims that only these can lead to genuine understanding. (An example of this attitude is the refusal of Leibniz and the Cartesians to admit that Newton had really *explained* anything.) Hume is adamant that in our new science we should not pretend to such "deeper" explanations. In particular, we should not give in to the temptation to try to explain *why* the principles of human nature our new method has revealed are the way they are: "we can give no reason for our most general and refined principles beside our experience of their reality." (T xxii) When he

[3] The relative importance of different influences on Hume from the direction of "natural philosophy" is a matter of controversy. The best-known interpretations (Noxon, Capaldi) assign the dominant role to Newton's influence; others (e.g., Barfoot) argue that Boyle's was greater or (Wright) that that of the mechanical philosophy, deriving from Descartes and Malebranche, perhaps greater still. Yet others (Jones) trace Hume's philosophy to quite other roots in classical thought. My discussion does not require taking sides on this question.

introduces his famous principles of association —the three "universal principles" which "guide" the operations of the imagination in uniting our ideas— Hume emphasizes that their reality requires no special proof beyond recognizing that their "effects are every where conspicuous." (T 13) He follows this claim immediately by reminding us that their "causes ... are mostly unknown, and must be resolv'd into *original* qualities of human nature, which I pretend not to explain" (*ibid.*, Hume's emphasis). It would be a mistake, however, to complain about this "impossibility of explaining ultimate principles" in the science of man, as it is "a defect common to it with all the other sciences, and all the arts, in which we can employ ourselves." (T xxii) Being sceptical about the possibility of answering the "deeper" questions of the metaphysician need not make one sceptical about the possibility of scientific knowledge.

Thus while there is certainly a sense in which Hume can be, as he so often is, said to be a sceptic, his scepticism should be understood as one about pretended supra-scientific metaphysical knowledge, rather than about scientific knowledge. It is this kind of scepticism that separates him from other philosophers before him, who conceived of philosophy as able to go beyond "mere" scientific knowledge and so able to provide a deeper and more certain knowledge of reality.[4] His extensive and devastating criticisms of the rationalists' attempts to deal with the threat of a scepticism that denies the possibility of scientific, indeed, any, knowledge, seemed to his contemporaries, and to many since, evidence that he shared their pre-occupation with that threat. Yet Hume is quite explicit in disclaiming such an interest and tells us clearly, in a variety of ways and contexts, that the main aim of his enquiries is something very different. An example is his admonition, in the opening paragraph of the section in the *Treatise* entitled "Of scepticism with regard to the senses," not to be concerned with the usual sceptical question about the existence of the external world:

> We may well ask, *What causes induce us to believe in the existence of body?* but 'tis in vain to ask, *Whether there be body or not?* That is a point, which we must take for granted in all our reasonings. (T 187)

That such an injunction should appear in this very section, nominally concerned with scepticism, is surely not an accident and should clinch the case that whatever Hume is doing, it is neither pressing, nor looking for—and failing to find—an answer, to the radical sceptical challenge. He tells us

[4] There are, in fact, other ways in which Hume can justly be said to be a sceptic, and recognizing them is important for a completely adequate interpretation of his overall philosophy, of which his science of the mind is only a part. But these are matters beyond my present topic. (For helpful discussions, see Fogelin (1985) and Fogelin (1993).)

explicitly what he *is* doing: "The subject, then, of our present enquiry is concerning the *causes* which induce us to believe in the existence of body... " (ibid., Hume's emphasis).

In sum, Hume *is* sceptical about various philosophical attempts at justifying our beliefs, especially when it comes to the most basic of these: in the external world, in the identity of our person, in causal connections, and the like, beliefs Hume thinks even a sceptic cannot seriously reject or live without. He often insists that it is just as well that nature has made sure that, in spite of all philosophy, we take these for granted, as without them "human nature would ... go to ruin" (T 225). But this recognition of our unreflective, instinctive and unavoidable acceptance of certain basic beliefs must not be confused with claiming to have a rational justification of them. It is philosophers' claims to the latter that are the targets of Hume's sceptical arguments, as are their pretensions to knowledge about the source of the principles a scientist of the mind can discover, and about the reasons why these principles are what they are.

I have noted that the method Hume urges for his new science consists of observation and experiment as much as does the method of natural philosophy. Yet it differs from the latter in some important respects. I shall discuss some of these differences later. But one must be noted right away: the impossibility in the new science of making experiments "purposely, with premeditation." Here Hume is thinking of the controlled experiments of the natural sciences, not of the thought experiments familiar in philosophy, which he himself often uses. In our new science we must "glean up our experiments... from a cautious observation of human life, and take them as they appear in the common course of the world, by men's behaviour in company, in affairs, and in their pleasures." (T xxiii) In spite of this, though, the science of man need "not be inferior in certainty... to any other of human comprehension." (ibid.)

2.

Hume, then, is quite explicit about the nature, status, and proper method, of the project he is undertaking. Yet his declarations have had remarkably little effect on the interpretation of his writings by champions and critics alike, from his day to ours. It is only very recently that some have begun to take him at his word and to see him as engaged in an inquiry at least continuous with what we today think of as the scientific study of the mind. But such recognition of continuity is still by no means universal, and it is an irony that the cognitive science of our day is sometimes explicitly contrasted with

Hume's science of man, rather than being recognized as the latter's descendant.[5]

One reason for this is the difference just noted in the two sciences' respective conceptions of observation and experiment. (I shall return to this.) Another, however, is the failure to appreciate the solutions he offers to the sceptical challenges he poses, solutions that contain, albeit in a form unfamiliar to many in recent cognitive science, at least the germs of some of the latter's central insights. I shall discuss examples of this failure later. First, however, let me offer a sketch of the main features of the picture of the mind that Hume, the anatomist, develops.

Hume's general answer to questions about how we come to have the various beliefs we have is that they are the product of a *non*-rational faculty. He labels this faculty variously as "instinct", "habit" or "custom", or "the imagination;" it is defined by a certain "propensity" to form ideas and beliefs. Some subtle differences behind this varying terminology notwithstanding, the main contrast is with reason, the faculty whose standards and operations some philosophers think can serve to provide an answer to the sceptics's challenge.[6]

The raw materials on which this non-rational faculty works and from which all mental life is constructed are impressions and their "faint copies," ideas, species both of the genus, perception: "All the perceptions of the human mind resolve themselves into two distinct kinds, which I shall call IMPRESSIONS and IDEAS." (T 1). Many of the most sceptical-sounding passages of the *Treatise* and of the first *Enquiry* are devoted to showing that our stock of these materials is more limited than philosophers have supposed. Hume shows us again and again that the impressions from which some putative idea posited by the metaphysician would have to derive are just not to be found in experience. What Hume calls his "first principle"—often labelled "the copy principle"—requires that all simple ideas be traceable back to simple impressions which they exactly resemble save for

[5] For an example of this, see Fodor (a). Contrast, however, Fodor (b), much more in sympathy with the view advocated here.

[6] Care has to be exercised with Hume's terminology here: while the chief contrast is between reason and the imagination, the latter is itself Janus-faced: "the general and more establish'd properties of the imagination," are to be sharply distinguished from "the trivial suggestion[s] of the fancy" (T 267). Which of these aspects of the imagination dominates in one's empirical reasoning ("reasonings concerning matters of fact") determines, for Hume, how reasonable one is. This makes sense only if we recognize that he uses terms such as 'reason' and 'reasoning' in two very different ways. In one sense, these terms refer to the kind of abstract reflective operation involved in the arbitrary comparison of ideas involved in what he calls "philosophical relations." In the other, they are labels for the kind of automatic, non-reflective, transitions that count as "natural relations," most important among them, causation. (For further discussion of these, see below.)

their lesser "forcefulness" or "vivacity." ("...all our simple ideas in their first appearance are deriv'd from simple impressions, which are correspondent to them, and which they exactly represent." (T 4)). But he does not deny the obvious, though remarkable, fact that from the rather limited stock of impressions that come my way, I am able to construct an edifice of beliefs that go far beyond those impressions and the ideas copied from them. First, my complex ideas are not confined to the complex impressions I have actually had: I can combine simple impressions in novel ways, into new complex ideas. (These, often called by Hume "fictions," may, but need not, give rise to belief, and they often do not, as with ideas of fiction in the usual sense. For Hume, though, not all "fictions" are fictions. See below.) Second, the course of my experience, the various patterns and regularities among the perceptions that make it up, is exploited by my mind in forming the beliefs I do. In both these ways, the mind must be conceived as essentially active: it is what it does with what it gets that matters, and it is this that Hume's science is designed to describe.

According to that science, the mind is led from one idea to another by three "principles of association": resemblance, contiguity, and cause and effect. These principles involve the mind's "taking notice" of certain properties of, and regularities among, its perceptions. Such taking notice need not be, and typically is not, conscious. What matters is that these properties and regularities be detected by the mind in a way that makes a difference to its subsequent operations and contents. Were it not for this active contribution on the mind's part, the mere presence of such properties and regularities would not be sufficient to explain the combinations and transitions among our ideas that actually occur, nor the genesis of the beliefs we actually form.

The remarkable regularities in the transitions we make from idea to idea and from (some) ideas to beliefs are the result of certain characteristics of the imagination, an ever-active (and sometimes over-active) non-rational faculty, the story of whose workings in large part constitutes Hume's scientific account of human nature. "Custom" or "habit" are Hume's usual short labels for these characteristics, among them a certain inertia that plays a role in Hume's explanations of some of the most remarkable, though often not even noticed, facts about the mind. These include the fact that in the absence of impressions from which the corresponding ideas could have been copied, we nonetheless come to believe that there are bodies and that we are the same person at one time as at another, even that we can "extend our identity beyond our memory" (T 262).

This inertia is one of the most important characteristics of the mind: "the imagination, when set into any train of thinking, is apt to continue, even when its object fails it, and like a galley put in motion by the oars, carries on in its course without any new impulse." (T 198) In his treatment of our ideas of space and time and of mathematical and geometrical reasoning, Hume appeals to it to explain how we generate an "imaginary standard of equality," notions of "perfection beyond what [our] faculties can judge of," and of "correction[s] beyond what we have instruments and art to make." (ibid.) Thus the "useless and incomprehensible fictions" of the mathematicians, who give exact definitions and demonstrations that have no experiential application. In this case, Hume's purpose is to expose these fictions as "absurd" (T 51-2). His recommendation is to resist the tendency and thus avoid the absurdity.[7] But this inertial tendency, automatic and non-reflective, is not limited to mathematical reasoning. It is ubiquitous: nothing, says Hume, is "more usual, than for the mind to proceed after this manner with any action, even after the reason has ceas'd, which first determined it to begin" (T 48). In the discussion of our belief in the existence of external objects, the same tendency is invoked in the interest of quite a different goal: that of explaining how we *cannot* avoid forming our "opinion of the continu'd existence of body:"

> Objects have a certain coherence even as they appear to our senses; but this coherence is much greater and more uniform, if we suppose the object to have a continu'd existence; and as the mind is once in the train of observing an uniformity among objects, it naturally continues, till it renders the uniformity as compleat as possible. (T 198)

Hume distinguishes between "principles which are permanent, irresistable, and universal" and those "which are changeable, weak, and irregular" (T 225). This distinction is essential to the double use by Hume of the inertial, extrapolating, tendency of the mind. When the tendency is guided by principles of the first sort, as it is in the formation of our fundamental common-sense beliefs, a recognition that this is so is what constitutes what Hume calls a "sceptical solution" to the sceptic's doubts, whether about the existence of an external world, about genuine causal connexions, or about personal identity. While Hume sometimes uses the term "fiction' to label a

[7] Some find Part II of the first book of the *Treatise* one of the most perplexing of the whole work, and its purposes are much disputed (See, for example, Anderson, Fogelin). But it is worth noting that at least at times Hume seems to be interested there in making an *anti*-sceptical argument. Having argued that the idea of extension as infinitely divisible is incoherent, he says: "Now 'tis certain we have an idea of extension; for otherwise why do we talk and reason concerning it? ... Here then is an idea of extension, which consist of parts or inferior ideas, that are perfectly indivisible: consequently this idea implies no contradiction: consequently 'tis possible for extension really to exist conformable to it: and consequently all the arguments employ'd against the possibility of mathematical points are mere scholastick quibbles, and unworthy of our attention" (T 32)

fundamental natural belief produced by this property of the mind, we must be careful not to be mislead by this into thinking of such a belief as somehow fanciful and arbitrary. Fictions of this sort are not optional: they are forced on us by our nature. But we can distinguish such fictions from those resulting from philosophical speculation floating free of common sense, such as, for example, the "feining" of a spiritual substance to explain the identity of a self over time. (I shall discuss Hume's own explanation of that identity later.)

In these cases of what we may call natural fictions, the mind's extrapolating tendency operates "in such an insensible manner as never to be taken notice of," and "the imagination can draw inferences from past experience, without reflecting on it; much more without forming any principle concerning it, or reasoning upon that principle." (T 102)

Hume adds that this tendency "may even in some measure be unknown to us." (ibid.) It is important to see that by this he means only that we have no introspective access to the processes in question. In forming a belief that one event is the cause of another, for example, I obviously do not consciously recall the previous instances of constant conjunction upon which my inference is based, using them as evidence for the inference. Not only does "the custom operate[s] before we have time for reflection," but "I am never conscious of any such operation," and in "giv[ing] preference to one set of arguments above another I do nothing but decide from my *feeling* concerning the superiority of their influence" (T 103, my emphasis). Thus it is that "all probable [that is, causal] reasoning is nothing but a species of sensation." (ibid.)

This distinction between reason (a reflective faculty purporting to make inferences on the basis of evidence) and the imagination (a non-reflective faculty that naturally moves from experience to belief) is fundamental to Hume's anatomy of the mind. To quote again from his discussion of our belief in external objects:

> "...our reason neither does, nor is it possible it ever shou'd, upon any supposition, give us an assurance of the continu'd and distinct existence of body. That opinion must be entirely owing to the IMAGINATION: which must now be the subject of our enquiry." (T 193)

Nor is this particular kind of belief unique in this respect: quite generally, "belief is more properly an act of the sensitive, than of the cogitative part of our natures" (T 183). When it comes to our most general and most fundamental beliefs (such as those in the existence of an external world, in our own identity, in causal relations), these are, therefore, quite impervious to the influence of reason, which can neither ground nor destroy them. "Cogitative part," means here our faculty of theoretical reasoning, at work when we construct demonstrations and philosophical arguments. There is, however, another sense of 'reasoning,', applicable to some of the natural and instinctive transitions we make from one perception one belief to another. Thus, for example, we are engaged in reasoning when we make a causal inference; indeed, that is what we primarily mean by 'reasoning' in ordinary, non-theoretical, contexts. ("... this inference is not only a true species of reasoning, but the strongest of all others" (T 97fn)). Hume calls this kind of reasoning "probable" or "experimental" and insists that we share it with infants, "nay, even brute beasts"—who presumably do not "cogitate" (*Enquiry,* 4.2). It is this latter kind of reasoning "on which the whole conduct of life depends, [and it] is nothing but a species of instinct or mechanical power, that acts in us unknown to ourselves" (ibid.).

3.

Hume's recommendation is to replace endless and fruitless "cogitating," in an attempt to give philosophical justifications of our beliefs, by the search for a scientific explanation of their origin. Doing so is what the "sceptical solution" of the sceptical challenge consists in (*Enquiry*, 5). It is to give up being a "metaphysician" and to become a scientist—an "anatomist"—of the mind. The resemblance between this recommendation and the so-called "naturalizing" programs common in recent philosophy of mind and epistemology is unmistakable. In these, too, the leading idea is to abandon a bankrupt a *priori* method in favour of an empirical, descriptive, one that holds out the promise of genuine progress. Many epistemologists have, in recent years, come to feel that arm-chair conceptual analysis is unlikely to tell us much about the real nature of human knowledge. (The uninspiring history of the so-called Gettier problem, involving increasingly arcane and artificial counter-examples to ever more byzantine definitions of knowledge is often taken as evidence of this.) Philosophers of mind, too, interested in understanding reasoning, perception, memory, language, and other mental phenomena, increasingly look to the new discipline (or constellation of disciplines) called "cognitive science" for illumination, rather than to inconclusive abstract philosophical analysis and argument.[8]

I have given examples elsewhere of how Hume's treatment of some of our fundamental concepts and natural beliefs anticipate, or are echoed in, cognitive science.[9] I want here to focus on the question I flagged earlier,

[8] For useful surveys of naturalization programmes in epistemology and in the philosophy of mind, see Goldman and Kornblith.

[9] Biro (d) and (e).

concerning the methods Hume uses in his science of the mind, both the one we have seen he recommends and a very different one we often find him actually using. I want to ask whether these different methods are compatible and, if so, how they are related to each other and to the methods of cognitive science. While the one he recommends is that of systematic every-day observation of others, the one that dominates Book I of the *Treatise* is one of first-person introspection. Neither is a method favoured by cognitive science, and it is worth asking three questions. First, are Hume's two methods compatible with one another? Second, in what ways do they differ from those of cognitive science? Third, do the methods Hume actually uses constitue a virtue or a limitation of his science?

We have already noted one striking difference between the method Hume recommends for his science and that of the "natural philosophy" on which, in other respects, that science is modelled: the difference between the use of controlled experiments and the observation of human behaviour as it occurs in the wild, as it were. (This difference is both reflected in, and obscured by, the differences between the ways Hume and we use the term 'experiment'.) In this regard, not only what we today call the "hard" sciences, but even much of modern social science (not all) fall on the non-Humean side of the divide. But in another way, a great deal of social science, by virtue of its being *social*, shares Hume's assumption that human nature cannot be studied in one human being, for it is found, indeed, in some of its most important aspects, is constituted, in the various relations people stand to each other. This aspect of Hume's science looms largest in the picture of the mind —the self, as he often calls it—found in the later parts of the *Treatise* and in other works, and it is in sharp contrast with the method prominent in Book I, with its more individualistic, atomistic picture. I shall return to it shortly. But first we need to note that the picture of the self drawn in Book I seems not at all to be based in the sort of "experiments" Hume recommends for the science of man. Rather, it involves a method common in empiricist philosophers before him and since, of looking to the meager materials provided by one's experience, from which one's ideas could be constructed or to which they could be reduced. The method this "looking" requires is one of observation, all right, but not of the things—"men's behaviour in company" and the like—recommended for the scientist of human nature, but of the sole, and curious, items directly accessible to consciousness: impressions and ideas. 'Observation' thus means, it seems, nothing more or less than *introspection*. No wonder, then, that most interpreters read the first book of the *Treatise* (and, since, as already noted, often not much else was read, the whole of Hume's philosophy) as part introspective psychology, part phenomenalistic reduction and part phenomenology—not to mention the scepticism![10].

The details of Hume's criticism of the substantial theory of the self favoured by rationalists, which occupies much of the section on personal identity in Book I, are too well known to require recounting. It cannot fail to be obvious to even the most superficial reader that those criticisms are based primarily on the evidence of introspection.[11] When Hume talks of "enter[ing] most intimately into what I call *myself*" (T 252) and of looking and not finding a perception of an enduring self distinct from "its" ever-changing impressions and ideas—which, if the substantial theory were true, he *should* find—it is difficult to see how any other method could be involved. Less obvious, and usually less noticed, is the pivotal role played by something like introspection in the positive account of the origin of one's idea of oneself as the same ("imperfectly" identical) over time that at least some commentators see him going on to develop.[12] This positive role of introspection is due to the importance of memory as the foundation of the natural relation, causation, the relation that ties together the diverse perceptions that constitute the complex object whose idea the idea of the self is. Causation is "to be consider'd ... as the source of personal identity," and "[H]ad we no memory, we never shou'd have any notion of causation, nor consequently of that chain of causes and effects which constitute our self or person."(T 262) Thus Hume's somewhat surprising conclusion: "...the memory not only discovers [the self's] identity, but also contributes to its production" (T 261)

But, of course, 'memory' in this context can mean only a special kind of perception, distinguishable from others by un-mediated internal observation, that is, phenomenologically. Perceptions of this sort refer to a time (putatively) earlier and to an experience (apparently) had at that earlier time, but do not imply anything about the actual existence of that time and

[10] The first of these readings is often a critical one, contrasting such psychology with real philosophy. (For a recent example, see Bennett). A combination of the second and the last was orthodoxy among analytic philosophers until recently; the third, again combined with elements of the sceptical interpretation, has been common in Europe (see, for example, Davie and Connon; for an attenuated phenomenologist reading, see Livingston); the last is too ubiquitous and too well-known to need documentation historically: it tempts just about everybody from Read until Kemp Smith (and even some today – see Stove, Fogelin).

[11] Primarily, though not exclusively. They also involve examining the way we use the language of identity and thereby providing an analysis of the concept of identity, resulting in the all-important distinction between perfect and imperfect identity. For detailed discussion, see Biro (a) and (b).

[12] Some do not see him as doing that at all (e.g., Penelhum); and it goes without saying that even those who do, hardly agree on what that account is. (Geach, Bricke, Robison, Traiger, McIntyre, Biro (a) and (b).)

that experience. So, the essence of the work memory does in tying together the perceptions constituting a self cannot lie in anything else than its phenomenologically, introspectively, available properties. The properties in question are obviously available in this way only to one mind or self, the one whose identity they account for.[13] Thus, even when Hume speculates on what one might find, were one able to "look into the breast of another," his very question underlines the essentially first-person character of both the information sought and the only possible method of seeking it.

The contrast with the overtly third-person, external, "observations" in which pursuing the new experimental method is said by Hume to consist, could hardly be greater. And this divergence re-appears at many places besides these discussions of the self, where it is easy to see that the nature of the problem Hume is struggling with—explaining the origin of one's idea of one's *own* identity[14]—demands a first-person approach. But in many other contexts, too, in which memory plays a central explanatory role—in grounding our ideas of time, of causation, of experience itself, as well as our belief in the existence of bodies—the first-person, introspective, method dominates Hume's discussion. I shall not give evidence for this general claim here, having done so elsewhere.[15] But to give just one example: to have an adequate (Hume calls it "just") idea of time, not only must we be presented with a succession of impressions such as Hume's "five notes play'd on a flute," but we must appreciate that succession (represent it to ourselves, as Kant, or some late-twentieth-century cognitivists might put it) *as* a succession. To quote William James on a closely related point: "A succession of feelings, in and of itself, is not a feeling of succession. And since, to our successive feelings, a feeling of their succession is added, that must be treated as an additional fact requiring its own elucidation."

The sort of fact involved here is, for both James and Hume, a phenomenological fact, and an appeal to facts of this sort seems to play a crucial role in many of the latter's explanations of how we come to have our fundamental ideas and beliefs. Such facts are of a very different sort than, and our access to them, and with it their epistemological status, is also very different from, those with which the science of man is supposed to be concerned. They are different also from the putative explanatory facts that science posits: we explain what we experience by postulating theoretical entities and processes that are not themselves part of that experience — explanatory entities to which we have no un-mediated introspective access. Chief among these are those propensities of the mind to which Hume

[13] For how to dispell the air of circularity about this, see Biro (b).

[14] For a defence of seeing Hume's treatment of personal identity in this light, see Biro (a).

[15] For details, see Biro (a), (e).

appeals again and again in his explanations of all our beliefs, both natural and philosophical. Familiar under the titles of "custom" and "habit," these natural, irresistible and unnoticed associating and extrapolating tendencies of the imagination *are* the scientific explanations Hume is seeking. They involve "experiments" and "observations" that consist not in peering *into* one's own mind in but looking *out* at the world, the method recommended by Hume for the anatomist of the mind.

Similar points could be made concerning almost any topic Hume treats in the *Treatise*.[16] He seems almost always to be engaged in two searches at once, one for the internal bed-rock, one for the properly objective, hence external, explanation. (And, as I just suggested, matters are further complicated by Hume's unabashed appeal to unobservables.) The two pursuits require him to use different methods; focusing on one to the exclusion of the other is one of the things that lead to the various one-sided and incompatible interpretations in which Hume scholarship abounds.

I shall come back shortly to the question of whether these various aims and methods are compatible. First, however, I want to re-emphasize the striking similarity between Hume's change of course from the speculative philosophical approaches to human nature in his time and that recommended by the various "naturalizing" programmes in epistemology and the philosophy of mind I mentioned earlier. Just as the former were meant by Hume to be a new departure promising real results and progress in the place of the idle —and worse— speculations of the metaphysicians, so the latter are, in large part, a reaction to what is seen as the triviality and inconclusiveness of arm-chair conceptual analysis and so-called "ordinary-language philosophy". As suggested earlier, the more and more *recherche* the examples and counter-examples of the Gettier industry, the less promise, it seemed, of genuine illumination about knowledge. In the philosophy of mind, the endless and seemingly pointless distinction- and idiom-mongering of some of the lesser lights of the ordinary-language school of the fifties and the sixties, along with disenchantment with various brands of behaviourism, led many to look, in something like the way Hume looked to Newton, to the emerging cognitive sciences as a new paradigm for how questions about the mind should be approached.

Many of the results of Hume's first efforts to begin a new science of the mind also have a clear echo in those being delivered by modern empirical studies of our cognitive processes. According to these, one of the most

[16] Compare his account of the difference between having an idea of a colour in terms of having a mastery of a colour concept and having a "just" idea of that colour in the sense of having phenomenal acquaintance with it (T 5); or his analysis of the term 'cause' with his account of the origin of the idea of necessary connection.

striking features of our cognitive capacities and performance, whether in perception, in linguistic processing, or in reasoning generally, is that the states, mechanisms and operations our best theories of them posit are often thought of as —to use current idiom— sub-doxastic, modular, and automatic. Sub-doxastic, because since their subject —the entity to which they are attributed— is not the cognizer himself, but some component sub-system we regard as the locus of the operation or process postulated to explain the cognitive function in question, we think of the states and processes involved as obtaining or taking place below the threshold of the cognizer's consciousness. For this reason, the subject is not a reliable source of information about these processes. Hence the preference in these studies for a third-person approach, rather than a first-person one, for laboratory experiments, such as reaction-time studies and the like, instead of introspection. (Compare Hume's advocacy, noted earlier, of "careful and exact experiments ... judiciously collected and compared," through "a cautious observation of human life.") Second, the processes of interest are modular, in that they are, in the overwhelming majority of cases, found to be task-specific, doing their work largely in isolation from each other and from the cognitive states we would attribute to the person taken as a whole. Thus the processes underlying one particular kind of cognitive capacity or performance rarely, if ever, interact with those of another, and their output is similarly independent of the output of other modules. (Think of the very common case of the different senses delivering different verdicts on the same object or event, with the need for a considerable amount of central processing to reconcile these). Finally, they are, for both these reasons, insensitive to the cognizer's beliefs —even if these are reflective and conscious, rather than merely tacit, and even if he makes an effort to bring them to bear on their workings. (Think of the robustness of perceptual illusions known to be illusions).[17] Hence the common characterization of many, if not most, processes posited to explain our cognitive capacities and performance as "cognitively impenetrable" or "informationally encapsulated."[18] We are not a long way from Hume's "instinct, or mechanical power, that acts in us unknown to ourselves."

[17] I have already mentioned Hume's recognition of our tendency to over-generalize. The same sort of inductive over-generalization has been found to be ubiquitous in our cognitive life, from linguistic processing (certainly in phonological and morphological processing, but even in syntax, e.g., with so-called "garden-path" sentences, and in prosody), to perception (of, for example, edges, or of motion), and general problem-solving and reasoning (as in making clearly fallacious probabilistic inferences). For details, see Ulman, Kahneman, Goldman.

[18] See Pylyshyn.

With all these similarities between Hume's new science and the even newer one of today, we should not overlook the important differences between them. We have already noted one: the difference in attitude to introspection as a method. Equally important is Hume's refusal to abandon those elements of the traditional framework that derive from common sense and our everyday practices, rather than from the rarefied and esoteric activities of philosophers *or* scientists. That is why, to return once again to his clear and explicit explanation of his method at the outset of his project, the experiments of the new scientist must consist of a "cautious observation of human life;" their objects must be taken "as they appear in the common course of the world." Hume therefore has a much more complex task than the modern cognitive scientist, or even the modern naturalizing philosopher. He must try to fit together into a coherent whole a number of elements that do not easily go together: introspection, scientific theorizing, conceptual analysis —and an ultimate allegiance to common sense and common language as the touchstone. This last desideratum complicates his task but, as I now want to suggest, it also yields the key to understanding the surprising extent to which he succeeds in it.

One of the striking features of the common-sense view of the world, and of human nature in particular, is that it is not solipsistic in the way that Hume's treatment of the self in Book I of the *Treatise* has appeared to many readers to be.[19] We naturally see ourselves as part of an external world, both physical and social. Not only is it idle to ask whether bodies exist (though it is not at all idle to ask how we come to believe that they do), it is equally idle to ask whether other people do. Surprising as this may sound to ears attuned to some of the traditional interpretations of Hume's account of the self, it will seem natural to those who remember that that account is not limited to, and is not completed in, the notorious discussion of Book I. The self of the later parts of the *Treatise*, social, engaged in action, seeing itself as so engaged, is not a different self from the lone introspective brooder of Book I, nor is it somehow—and questionably—derived from the latter by some inference. It is, rather, the very same self considered in a different light. Indeed, this later self is the real self, so to say, the one we all take ourselves to be. And that self depends for its very being and identity on its relations to others like it. Hume's emphasis on the mechanism of sympathy so prominent in the later books would not make any sense were this not so.

[19] I have myself emphasized the first-person and foundational character of that treatment (Biro (a)). But it is important to recognize that Hume's aim in Book I is descriptive and explanatory, rather than justificatory: it is to give a psychological account of how one's beliefs (including one's belief in one's own identity) arise, rather than an epistemological grounding of them. Part of Hume's message is that first-person materials are inadequate for the latter. (See Biro (b) and below.)

I have put the point in the strong, even extreme, way I just have —the self depends for its very being and identity on others— to draw attention to another striking anticipation of twentieth-century philosophical developments by Hume, this time in a very different neck of the philosophical wood. That the self is, in a strong sense, a social product is a common-place in much of the still vital philosophical tradition that derives from German and French thinkers of the nineteenth and twentieth centuries. Social-construction theories of identity, of personhood, of ethnicity, of gender, of —you name it, are all over the place, some to be taken seriously, some... well... Whatever their quality, these theories all rely essentially on the insight elegantly expressed by Hume in saying that "the minds of men are as mirrors to one another" (T 365), and systematically followed up by him in the later books of the *Treatise* (and elsewhere) to yield what could be defended as the most subtle, most comprehensive, most coherent, and most realistic, moral psychology and moral philosophy since Aristotle.[20] A full justification of this claim is, of course, well beyond the scope of the present paper. For now it must serve merely to take us back to the question I postponed a little way back: Can these different aspects of Hume's inquiries, and the different methods they require him to use, be fitted into one coherent whole, can they be seen as complementary parts of a coherent scientific-*cum*-philosophical project? I suggest that the answer to at least the latter question may be "Yes!," even if we have to concede that Hume himself perhaps does not quite succeed in fully knitting the different strands together.

To make out the correctness of an affirmative answer would take much evidence and argument. I have enough of neither, nor the space to give what I do have. So, I shall close by giving what is no more than the barest outline of an account that, if made out, would yield such a positive answer.

The key to the account I have already gestured at: it lies in placing Hume's *theories* of the nature and workings of the mind on the right side of the distinction between the justificatory project of traditional epistemology and the descriptive one of cognitive (and moral) psychology. It is important to emphasize that this extends even to the first of the two aims I attributed to Hume, namely, to that part of his explanation of the genesis of our beliefs in the service of which he employs the first-person, introspective, phenomenological method. As we have seen, the explanation is not complete without its complementary third-person, experimental, scientific element. But in both components Hume's goal is to discover by one or another kind of observation how the mind actually works, in contrast to others who either

[20] Perhaps one can see so-called social externalism of the kind advocated by Burge as an expression of one aspect of the same insight.

speculate on, or stipulate, how it *must* work. (Neither, as I have argued, is it his goal to show that we cannot know how it works.) And, while I cannot go into the matter here, I want to note at least that I believe it can be shown conclusively that this descriptive thrust of Hume's project does not preclude his giving an account of immanent normativity.[21]

From this perspective, we can, I think, say that while there is work for both methods to do, they are perhaps, in the end, neither as different nor as independent of each other as they may at first sight seem. The link between them lies in the role of language: in the non-solipsistic contents of the terms we must rely on even when we adopt a first-person method. This is, of course, a long story, one to which I can no more than point.[22] But the basic idea is that even when turning inward, what one finds are items whose very identity depends on their representational content, which is, in turn, inseparable from the meaning of the words one uses to express it. Hume's theory of meaning, so far from being one on which, as it is often alleged, the meaningfulness of words derives from mental items with no representational or semantic properties, is just the opposite: it is the public meanings of the words that individuate the mental items to which they (purport to) point. Thus it is that he can allow that even the blind can be said to possess ideas —concepts, as we would say—of colours in virtue of their ability to use colour terms. Hume's phenomenology may be seen as what Dennett has called "hetero-phenomenology," in contra-distinction to "auto-phenomenology," as turning, before the reader's eyes, into something like what Austin must have had in mind in speaking as he did of "linguistic phenomenology." And should not be surprised if before long cognitive scientists themselves come to question the assumption, so far dominant in their field, that a science of cognition requires an internalist, individualist, approach of the sort *one* strand in Hume's method exemplifies.

As I have said, all this is sketchy, perhaps hopelessly so. Still, it may give a feel for how Hume's new —at his time— insights about the mind quite uncannily anticipate those of recent cognitive science. While his idiom is, of course, not ours, while there are real differences between the method he recommends and the one he actually uses, as well as between these and those we favour, (and while his philosophical concerns go beyond the scientific study of the mind, something I have had to put aside in this

[21] Among those who *have* shown this, in various ways, are Ardal, Norton, Penelhum, and Swain.

[22] Some of the elements of the story have been provided by writers like Ardal, in his account of Hume's conventionalist account of language, Livingston, in his account of Hume's historical —and therefore social— theory of meaning, and Baier and others who have emphasized the essentially social picture Hume ultimately gives of the self.)

discussion), he may nonetheless be said with more justice than any philosopher before him to be the philosophical father of cognitive science.

REFERENCES

Anderson, R.F., *Hume's First Principles* (Lincoln: University of Nebraska Press. 1966).

Ardal, P., "Convention and Value," in *David Hume: Bicentenary Papers,* G.P. Morice ed., (Edinburgh: University of Edinburgh Press, 1977).

Austin, J.L., "A Plea for Excuses," in *Philosophical Papers* (Oxford: Oxford University Press, 1961).

Baier, A., *A Progress of Sentiments: Reflections on Hume's Treatise* (Cambridge: Harvard University Press, 1991).

Barfoot, M. "Hume and the Culture of Science in Early Eighteenth-Century Britain."

Bennett, J., *Locke, Berkeley, Hume: Central Themes* (Oxford: Oxford University Press, 1971).

Biro, J.I. (a),"Hume on Self-Identity and Memory," *The Review of Metaphysics* 30, 1976.

- (b),"Hume's difficulties with the self," *Hume Studies* V, 1979.

- (c),"Description and Explanation in Hume's Science of Man," *Transactions of the Fifth International Congress on the Enlightenment,* (New York: The Voltaire Foundation, 1979).

- (d), "Hume and cognitive science," *History of Philosophy Quarterly* 2, 1985.

- (e), "Hume's New Science of the Mind," in Norton (b).

- (f), "Memory, Mind, and Society," unpublished mss.

Bricke, J. (a), "Hume on Self-Identity, Memory and Causality," in Morice.

- (b) *Hume's Philosophy of Mind* (Princeton: Princeton University Press).

Burge, T., "Individualism and the Mental," *Midwest Studies in Philosophy*, 1979.

Broughton, J., "What does the scientist of man observe?" *Hume Studies*, vol. XVIII, no.2, 1992.

Capaldi, N., *David Hume: The Newtonian Philosopher* (Boston: Twayne Publishers, 1975).

Connon, R.W., "The Naturalism of Hume Revisited," in Norton (a).

Dennett, D., *Brainstorms* (Cambridge: MIT Press, 1978).

Davie, G., "Edmund Husserl and 'the as yet, in its most important respect, unrecognized greatness of Hume'," in Morice.

Fodor, J., (a) "Mental Representation: An Introduction," in Rescher.

- (b) *Humean Variations* (mss.).

Fogelin, R.J., (a) *Hume's Scepticism in the Treatise of Human Nature* (London: Routledge and Kegan Paul, 1985)

- (b) "Hume's scepticism," in Norton (b).

Garrett, D., *Cognition and Commitment in Hume's Philosophy* (New York: Oxford University Press, 1997).

Goldman, A., *Epistemology and Cognition* (Cambridge: Harvard University Press, 1987).

Hume, D., (a) *A Treatise of Human Nature*, L.A. Selby-Bigge, ed., rev. by P.H. Niddich (Oxford: Clarendon Press, 1987).

- (b) *An Enquiry Concerning Human Understanding*, L.A.Selby-Bigge, ed., (Oxford: Clarendon Press, 1970).

James, W., *The Principles of Psychology* (New York: Dover, 1950).

Kahnemann, D et al. (eds.), *Judgments under Uncertainty: Heuristics and Biases* (Cambridge: Harvard University Press, 1982).

Kornblith, H. (ed.), *Naturalized Epistemology* (Cambridge: Harvard University Press, 1985).
Livingston, D. (a), "Hume's Historical Theory of Meaning," in Livingston and King.
- (b), *Hume's Philosophy of Common Life* (Chicago: University of Chicago Press, 1984).
Livingston, D. and King, J. (eds.), *Hume: A Re-evaluation* (New York: Fordham University Press, 1976).
Monteiro, J-P., "Hume's Conception of Science," *Journal of the History of Philosophy* 19, 1981.
Morice, G.P. (ed.), *David Hume: Bicentenary Papers* (Edinburgh: University of Edinburgh Press, 1977).
Norton, D.F. (ed.), *The Cambridge Companion to Hume* (Cambridge: Cambridge University Press, 1993).
Norton, D.F. et al. (eds.), *McGill Hume Studies* (San Diego: Austin Hill Press, 1979).
Noxon, J., *Hume's Philosophical Development* (Oxford: Clarendon Press, 1973).
Owen, D., *Hume's Reason* (Oxford: Oxford University Press, 1999).
Penelhum, T., (a) "Hume on Personal Identity," *The Philosophical Review* 54, 1955.
- (b) "Hume's Theory of the Self Revisited," *Dialogue* 14, 1975.
- (c) *Hume* (London: Macmillan), 1975.
Pylyshyn, Z., *Computation and Cognition: Toward a Foundation for Cognitive Science* (Cambridge: Bradford/MIT Press, 1984).
Rescher, N. (ed.), *Scientific Inquiry in Philosophical Perspective* (Washington: University Press of America, 1987).
Robison, W.L., (a) "Hume on Personal Indentity," *Journal of the History of Philosophy* 12, 1974.
- (b) "In Defense of Hume's Appendix," in Norton (a).
Rosenberg, A., "Hume and the Philosophy of Science," in Norton (b).
Shope, R.K., *The Analysis of Knowledge: a Decade of Research* (Princeton: Princeton University Press, 1983).
Smith, N.K., *The Philosophy of David Hume* (London: Macmillan, 1949).
Stove, D.C., "The Nature of Hume's Skepticism," in Norton (a).
Stroud, B., (a) *Hume* (London: Routledge and Kegan Paul, 1977).
- (b) "Hume's Scepticism: Natural Instincts and Philosophical Reflection," *Philosophical Topics* 19.
Swain, C. "Being Sure of One's Self: Hume on Personal Identity," *Hume Studies,* 1991.
Ullman, S., *The Interpretation of Visual Motion* (Cambridge: Harvard University Press, 1979).
Winkler, K., "The New Hume," *Philosophical Review* 50, 1991.
Wright, J., *The Skeptical Realism of David Hume*, (Manchester: Manchester University Press, 1983).

Chapter 2

TRUTH AND MEANING

The Role of Truth in the Semantics of Propositional Attitude Ascriptions

Scott Soames
Princeton University

In contemporary philosophy, theories of truth take two main forms. One form aims to provide an analysis of truth that illuminates the content of claims that x is true or not true, that explains the uses to which the notion of truth is legitimately put, and that can be employed to diagnose and resolve paradoxes involving truth, such as the liar. The other main form taken by contemporary theories of truth is found in semantics, where theories of truth are used to interpret, or give the meanings of, sentences. When we use theories of truth to play this role, we are not attempting to provide an analysis of truth, or to specify the content of the truth predicate. Rather, we take truth for granted, and use it to illuminate the meanings of sentences by giving their truth conditions. Here it is usually taken for granted either that the meaning of a sentence consists of nothing more than its truth conditions, or that truth conditions constitute a central component of that meaning. As a result, theories of meaning are sometimes taken to be nothing more than theories of truth, while at other times theories of truth are viewed as central subcomponents of theories of meaning.

Today I will talk about the role of truth in theories of meaning. My focus will be on propositional attitude ascriptions, and the problems they pose for truth-conditional theories. I will argue that contemporary theories that aspire to interpret, or specify the meanings of, such ascriptions, simply by deriving their truth conditions, are doomed to failure. Next, I will argue that although it remains possible for certain truth-conditional theories to provide some information about the meanings of propositional attitude ascriptions, there

K. Korta and J. M. Larrazabal (eds.), Truth, Rationality, Cognition, and Music, 21-44.

are serious difficulties in extracting substantive testable consequences from even these theories.

In both cases my argument will focus on the role played by canonical theorems of the form ***'S' is true in L iff P***, ***'S' is true in L with respect to a possible world w iff in w P***, or , ***'S' is true in L with respect to a context of utterance c and possible world w iff in w*** P_c (where the specification of P_c often depends on c).[1] Standardly, claims of this sort are thought of as making up the chief empirical content of theories of meaning, and as providing the basis for the interpretation of object-language sentences. In order to play these roles, such claims must themselves be capable of being understood and evaluated independently of the theory. For this, two things are required: (i) We must understand in advance of the theory what it is to say of a sentence of L that it is true, and often we need to have some idea of what to do to provide evidence that this or that particular sentence is, or is not, true. (ii) We must understand and be able to evaluate, independently of the theory, the right-hand sides of the biconditionals stating the truth conditions of object-language sentences generated by the theory. In the case of propositional attitude ascriptions, I will show that leading contemporary theories of truth and meaning fail to satisfy (ii). This is no accident; the constraint is very difficult to satisfy.

1. A PROBLEM FOR DAVIDSONIAN SEMANTICS OF PROPOSITIONAL ATTITUDE ASCRIPTIONS

I begin with one of the latest, and most sophisticated, applications of the Davidsonian program to the semantics of propositional attitude ascriptions – the analysis given by Richard[2] Larson and Peter Ludlow in their paper, "Interpreted Logical Forms." The analysis is Davidsonian in two respects. First, they accept Davidson's contention that theories of truth can play the role of theories of meaning. Second, the analysis of propositional attitude ascriptions they offer is a recognizable descendant of Davidson's own early analysis in "On Saying That."[3] In that paper Davidson viewed attitude ascriptions as reporting relations between agents and particular utterances. Since this view of utterances as objects of the attitudes gives rise to problems, Larson and Ludlow update the account by viewing attitude

[1] Here, and throughout this paper, I use boldface italics to play the role of corner quotes.

[2] *Synthese*, 95, 1993, pp. 305 - 356, reprinted in *Readings in the Philosophy of Language*, Peter Ludlow, ed., (Cambridge Mass.: MIT Press, 1997). Citations will be from the reprinted version.

[3] Donald Davidson, "On Saying That," *Synthese*, 19, 1968-9, pp. 130-46.

ascriptions as reporting relations between agents and what they call "interpreted logical forms." An interpreted logical form for a sentence S is a syntactic representation of S, in particular a phrase marker the terminal nodes of which are the individual words of S paired with n-tuples of objects thought of as "semantic values." The non-terminal nodes of the phrase marker are associated with values related to the semantic values of their daughter nodes.

A pictorial representation of the interpreted logical form associated by Larson and Ludlow's semantic theory with the sentence (1a) is given by (1b).

1a. John speaks Spanish.

b.

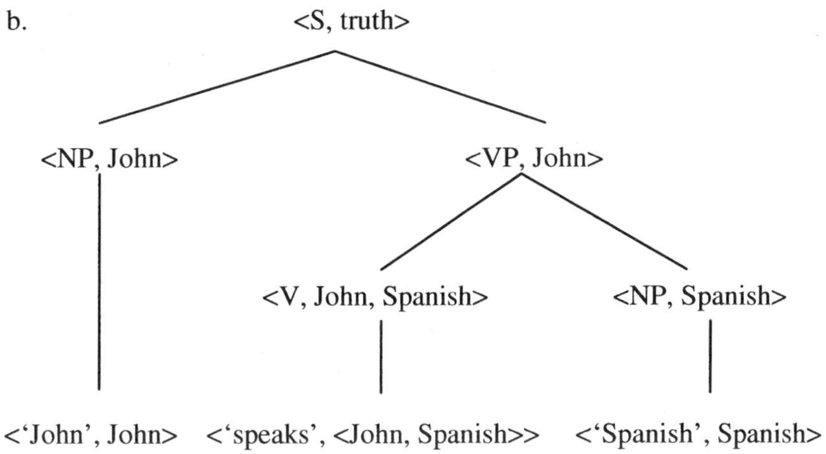

The proposed semantic theory works in the following way: we are given as input to the theory grammatically analyzed sentences of the object language in effect sentences with explicit syntactic representations of their hierarchical organization into constituents and subconstituents. The semantic theory is a Davidson-style theory of the material truth conditions of grammatically analyzed sentences.

Rough examples of the rules used to interpret (1a) are given in (2).

2a. A sentence of the form [$_S$ NP VP] is true iff there is an object x which is a semantic value of both NP and VP.

b. x is a semantic value of [$_{NP}$ 'John'] iff x = John.

c. x is a semantic value of [$_{NP}$ 'Spanish'] iff x = Spanish.

d. x is a semantic value of [$_{VP}$ V NP] iff there is an object y such that <x,y> is a semantic value of V and y is a semantic value of NP.

e. <x,y> is a semantic value of [$_{V}$ 'speaks'] iff x speaks y.

These rules are used to derive the statement of the truth conditions of (1a) given in (3).

3. The (grammatically analyzed sentence) '[$_{S}$ [$_{NP}$ John] [$_{VP}$ [$_{V}$ speaks] [$_{NP}$ Spanish]]]' is true iff there is an object x such that x = John and there is an object y such that y = Spanish and x speaks y—in effect, iff John speaks Spanish.

To get the interpreted logical form (1b) that is assigned by the theory to sentence (1a), one starts with the uninterpreted logical form (4), which is given by the syntax and which serves as the input to the semantic theory.

4.

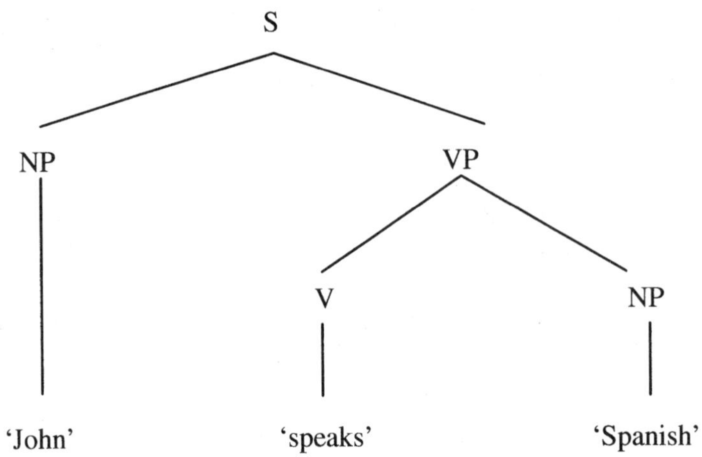

One then attaches a putative semantic value to each of the nodes to get the interpreted logical form.[4] The value one attaches to the S-node is always truth, and the value one attaches to any other node is the unique object which

[4] Not necessarily its real semantic value. For example the sentence may not, in fact, be true, and the pair <John, Spanish> may not really be one of the pairs to which 'speaks' applies.

can be proved to be a semantic value of that node from the assumption that the sentence is true, using the semantic rules for the sentence. So, in our example, the rules (2b) and (2c) give us directly that the semantic value of the NP-node dominating 'John' is the man John, and the semantic value of the NP-node dominating 'Spanish' is the Spanish language. From this together with (2a) and the assumption that the sentence is true we derive that the man John is a semantic value of the VP-node as well. Combining this with the rule (2d) for verb phrases gives us the ordered pair of <John, Spanish> as a semantic value of the V-node dominating 'speaks'. The end result is the interpreted logical form (1b).

According to Larson and Ludlow, such interpreted logical forms are objects of the attitudes. The rules for the attitude verbs 'assert' and 'believe' are given in (5).

5a. <x,y> is a semantic value of 'asserts' iff x asserts y.

b. <x,y> is a semantic value of 'believes' iff x believes y.

The rule that tells us that the objects of belief and assertion ascriptions are the interpreted logical forms of their complement clauses is (6).

6. x is a semantic value of [$_{VP}$ V S] —i.e. of a verb phrase consisting of a propositional attitude verb followed by a complement clause— iff there is a y such that <x,y> is a semantic value of V and y = the interpreted logical form assigned to S (in accord with the above process).

By putting all of these rules together one can derive theoretical statements of the truth conditions propositional attitude ascriptions like (7).

7. Mary asserts / believes that John speaks Spanish.

The theorem of the semantic theory interpreting this sentence will have roughly the form (8).

8. The (grammatically analyzed sentence) '[$_S$ [$_{NP}$ Mary] [$_{VP}$ [$_V$ asserts / believes] [$_S$ that John speaks Spanish]]]' is true iff Mary asserts / believes ...... (Fill in the dots with a description of (1b).)

If we were to spell out in words the intended content of the theorem (8) that is derivable from the theory, it would be something like (9).

9. The (grammatically analyzed sentence) ‘[$_S$ [$_{NP}$ Mary] [$_{VP}$ [$_V$ asserts / believes] [$_S$ that John speaks Spanish]]]’ is true iff Mary asserts / believes the interpreted phrase marker the root node of which is the pair <'S', truth>, which dominates a pair of nodes <'NP', John> and <'VP', John>, where <'NP', John> dominates <‘John’, John>, and <'VP', John> dominates a pair of nodes <'V', <John, Spanish>>, and <'NP', Spanish>, with the first of these nodes dominating <‘speaks’, <John, Spanish>> and the second dominating <'Spanish', Spanish>.

When spelled out in full detail, this T-sentence, allegedly interpreting the attitude ascription (7) is quite a mouthful. Certainly no truth theory issuing such theorems is homophonic, a fact noted by Larson and Ludlow with heroic understatement in a footnote.

More troubling is the fact that the T-sentences are themselves mysterious, and hard to interpret. At this point it is essential that one not get caught up in technicalities, but rather step back and ask what is going on. The task we are engaged in is to give a semantic theory for propositional attitude ascriptions. According to Larson and Ludlow, this means giving a Davidsonian theory of the truth conditions of those ascriptions. The most crucial aspect of this theory is that it is supposed to issue in theorems of the form

10. S is true (in L) iff .....

that interpret the object language sentence S by giving its truth conditions. Suppose, however, that a particular theory of this sort generates theorems of the form (10) in which the clause filling in the dots on the right-hand side of the biconditional expresses a theoretical claim that we have no pretheoretic understanding of, nor any ability to directly test or evaluate. In such a situation the semantic theory cannot be said to have accomplished the fundamental task of giving interpretive truth conditions for object language sentences.

That is precisely the situation that Larson and Ludlow's semantic analysis of attitude ascriptions confronts us with. Suppose that someone walked up to you, and assertively uttered, out of the blue, the right-hand side of (9). That is, suppose the person simply said these words "Mary believes the interpreted phrase marker the root node of which is the pair <'S', truth>, which dominates ...". Would you understand the person’s remark, or know how to determine whether it was true? Surely not. Prior to being convinced of the correctness of some contentious theory, we have no idea what it would mean to say that a person believes one of these unfamiliar objects. Thus, until we are given an explanation of what it might mean to believe or assert what Larson and Ludlow call interpreted logical forms, accompanied by persuasive empirical evidence that we really do believe and assert these things, we won't have any grip on the right-hand sides of T-sentences like (9).

The problem is that Larson and Ludlow provide no such explanation. It is, of course, tempting when reading them to rely on the fortuitous fact that we already understand the attitude ascriptions mentioned on the left-hand sides of the T-sentences. Given this understanding, we might be tempted to use the T-sentences to interpret the theoretical claims on their right-hand sides. But this is illegitimate, from the point of view of the overall Davidsonian framework that Larson and Ludlow adopt. The information provided by the theory about the full range of theorems like (9) is supposed to allow someone with no antecedent knowledge of the object-language to interpret, and come to understand, the object-language sentences on the left. It is impossible for the propositions expressed by the T-sentences produced by Larson and Ludlow to play this role if the meanings of the object-language sentences are going to be presupposed, and used to interpret the theoretical claims on the right-hand side.

This problem for Larson and Ludlow arises directly from their commitment to a Davidsonian conception of semantic theories. It is a central feature of the Davidson program that no meanings are assigned to sentences, and no theorems stating what sentences mean are derivable from Davidsonian semantic theories. Because of this, some independent justification is needed to support the claim that a Davidsonian truth theory for a language may play the role of a theory of meaning. The difficulty for Larson and Ludlow is that obscure and theoretically loaded theorems like (9) cannot play the justificatory role standardly required by Davidsonians in order for their theories of truth conditions to count as theories of meaning.

Davidson himself has made two main attempts to justify the claim that a theory of truth for a language may play the role of a theory of meaning. According to the first, knowledge of that which is stated by a Davidsonian truth theory is claimed to be sufficient in order for an agent to learn, and become a competent speaker of, a language. The idea here is that any theory that tells us everything we need to know in order to understand a language may be counted as specifying all essential facts about meaning, even if it does not generate theorems that explicitly state the meanings of individual sentences. In "Truth and Meaning"(1967) and "Radical Interpretation" (1973),[5] Davidson claimed that theories of truth of the sort he advocated

[5] Donald Davidson, "Truth and Meaning," *Synthese*, vol. 17, 1967, pp. 304-323, and "Radical Interpretation," *Dialectica*, vol. 27, 1973, pp. 313-328.

satisfy this condition, despite the fact that the T-sentences, ***'s' is true in L iff p***, entailed by such theories were **not,** at that time, required by Davidson to be translational in the sense of pairing the object-language sentence s with a metalanguage sentence p that is a translation of s. When, in 1976, it was proven by J. A. Foster[6] that Davidsonian truth theories do not, in fact, provide information knowledge of which is sufficient for understanding a language, Davidson's first attempt to justify the claim that truth theories may play the role of theories of meaning collapsed, and he made a second attempt to justify this claim.[7]

According to this second attempt at justification, in order to qualify as theories of meaning, theories of truth must entail translational T-sentences that pair object language sentences mentioned on the left with metalanguage paraphrases used on the right. The new idea, roughly put, was that knowing both (i) that which is stated by a theory of truth T, and (ii) that T yields a translational T-sentence for each object language sentence, should be sufficient for understanding the language. Although this attempted justification has itself been shown to suffer from serious problems,[8] it cannot be appealed to by Larson and Ludlow in any case. The metalanguage sentences that occur on the right-hand sides of T-sentences like (9) are about as far from translational paraphrases of the attitude ascriptions on the left as one can get. Since these T-sentences are not translational, Larson and Ludlow cannot invoke any version of Davidson's second attempt to justify the claim that their truth theories should count as a theories of meaning. More generally, being Davidsonians they are faced with a dilemma. They are confronted with two problems: (i) justifying the claim that theories of truth may qualify as theories of meaning, and (ii) giving the truth conditions of attitude ascriptions. Their dilemma is that their solution to (ii) conflicts with standard Davidsonian approaches to (i).

[6] J. A. Foster, , "Meaning and Truth Theory," in *Truth and Meaning*, Gareth Evans and John McDowell, eds., (Oxford: Oxford University Press, 1976).

[7] See Davidson, "Reply to Foster," in *Truth and Meaning*.

[8] For a detailed discussion of the problems with both attempts, see Scott Soames, "Truth, Meaning, and Understanding," *Philosophical Studies*, vol. 65, 1992, pp. 17-35.

2. THE EMPIRICAL CONTENT OF SEMANTIC THEORIES: A BROADER DIFFICULTY[9]

In my opinion, this dilemma is intractable from the Davidsonian perspective. It is important to note that it does **not** generalize to other semantic approaches. In particular, it does not generalize to semantic theories that assign certain objects – often called propositions – to sentences as their semantic contents in contexts, and that identify meanings of sentences with functions from contexts to contents. Since these more complex theories make explicit claims about what the meanings of the sentences are, they do not have to satisfy some extra condition – like the condition that knowledge of that which is stated by the theory must be sufficient to make one a competent speaker of the language – in order to qualify as theories of meaning. In this, they have an advantage over the Davidsonian program. Of course, they also incur an obligation to specify the empirical evidence that can be adduced to support their theoretically loaded claims about meaning. Standardly, the central way in which this obligation is discharged is by the assignment of truth conditions – first to propositions, and then, derivatively, to sentences that express those propositions relative to contexts. Thus, claims about the truth conditions of sentences relative to contexts come make up the central core of potentially testable empirical content not just of Davidsonian theories, but of richer and more complex semantic theories as well.

With this in mind, let us return to the difficulties we have raised for Larson and Ludlow's account of propositional attitude ascriptions. One problem was that since the right-hand sides of the T-sentences generated by the theory are neither translational paraphrases of the attitude ascriptions mentioned on the left, nor claims that we pre-theoretically understand, explicit acceptance of these T-sentences and the truth theories that generate them will not suffice to understand the object language attitude ascriptions; hence the Davidsonian justifications for taking such truth theories to be theories of meaning will be defeated. However, there is also a second problem that goes well beyond the special requirements of the Davidsonian program. This second problem is that we don't even know what truth conditions Larson and Ludlow have assigned to attitude ascriptions, because we don't know how to evaluate the theoretical claims made on the right-hand sides of theorems like (9). Unless we are given a way of extracting

[9] The discussion of Larson and Ludlow in both this and the previous section is closely based on the discussion of their theory in the first few pages of chapter 7 of my book, *Beyond Rigidity: The Unfinished Semantic Agenda of Naming and Necessity*, (New York: Oxford University Press, 2002).

testable empirical content from these theorems, we have no way of determining whether the claims they make about truth conditions are correct.

This problem does generalize to other semantic approaches. Many semantic theories attempt to state the truth conditions of attitude ascriptions in terms of the relations agents bear to the objects of those attitudes. Different theories characterize these objects differently – e.g. as interpreted logical forms, structured propositions, or sets of possible worlds. However, no matter what the characterization, each theory faces the task of explaining what it is to bear the relevant attitude relation to the objects identified by the theory. It is only by giving such explanations that semantic theories can make empirically testable claims about the truth conditions of attitude ascriptions. What has not, I think, been adequately appreciated is how difficult it can be to extract such predictions from semantic theories. Since this point applies with special force to the theory of Larson and Ludlow, I will begin by using it to illustrate the more general point.

For this purpose I want you now to put aside the question of whether Davidsonian theories could ever qualify as theories of **meaning**, and concentrate instead on the more limited task of evaluating the truth or falsity of the theorems actually generated by such theories. In the case of Larson and Ludlow, we need some set of additional principles that will allow us to derive testable empirical predictions from theorems like (9). Since these theorems link claims about the truth of object language sentences to highly theoretical claims of which we have no pretheoretic understanding, something further must be said before we can even determine what truth conditions have been attributed to these sentences.

First, observe that Larson's and Ludlow's analysis claims that the attitude ascription (7) is true iff Mary asserts or believes the interpreted logical form (1b). If the theory also told us that the interpreted logical form (1b) were true iff John speaks Spanish, then we could put these two results together to derive the conditionals (11a) and (11b).

11a. If the (grammatically analyzed) sentence '[$_S$ [$_{NP}$ Mary] [$_{VP}$ [$_V$ asserts / believes] [$_S$ that John speaks Spanish]]]' is true, then Mary asserts / believes something that is true iff John speaks Spanish.

11b. If the (grammatically analyzed) sentence '[$_S$ [$_{NP}$ Mary] [$_{VP}$ [$_V$ asserts / believes] [$_S$ that John speaks Spanish]]]' is not true, then there is something that is true iff John speaks Spanish which is such that Mary doesn't assert / believe it.

Results like these would at least provide some correct and testable information about the truth conditions of the attitude ascription.

As it stands, the theory does not provide an account of the truth conditions of interpreted logical forms like (1b), and so it does not allow us to derive (11a,b). However this could easily be fixed. The theory does give an account of the truth conditions of what one might call "uninterpreted logical forms" like (4); and this account could be extended to provide a treatment of the truth conditions of interpreted logical forms like (1b) that would allow us to derive informative theorems like (11a) and (11b).

Although this is progress, we are not out of the woods. Since Larson's and Ludlow's semantic theory provides only material truth conditions, the information provided by these theorems is minimal. In effect, (11a) tells us that if the attitude ascription (7) is true, then there is some p such that Mary asserts / believes p, and p has the same truth value in the actual world as the claim that John speaks Spanish; (11b) tells us that if (7) is not true, then there is some p such that Mary doesn't assert / believe p and p has the same truth value in the actual world as the claim that John speaks Spanish. Note, in addition to being quite weak, these conditionals cannot be turned into biconditionals. Thus, (11a) and (11b) do not provide necessary and sufficient truth conditions for the attitude ascription (7), even in the minimal sense standardly demanded of Davidsonian truth theories. Although (11a) and (11b) do provide some clear, testable information about the truth conditions of (7), this information falls well short of what we need to evaluate theorems like (9).

If the theory could be transformed into one that assigns sentences and interpreted logical forms truth conditions relative to different possible worlds, then the information provided by the theory would be substantially enhanced. Although this goes substantially beyond Larson and Ludlow's actual discussion, it too seems as if it is the sort of thing that could in principle be done. If it were done, one could generate strengthened versions of (11) along the lines of (12a) and (12b).

12a. If the (grammatically analyzed) sentence '$[_S$ $[_{NP}$ Mary] $[_{VP}$ $[_V$ asserts / believes] $[_S$ that John speaks Spanish]]]' is true with respect to a possible world w, then there is something p that is true with respect to an arbitrary world w* iff in w* John speaks Spanish and, in w Mary asserts / believes p.

12b. If the (grammatically analyzed) sentence '$[_S$ $[_{NP}$ Mary] $[_{VP}$ $[_V$ asserts / believes] $[_S$ that John speaks Spanish]]]' is not true with respect to a possible world w, then there is something p that is true with respect to an arbitrary world w* iff in w* John speaks Spanish and, in w Mary does not assert / believe p.

These theorems are more informative than (11a,b). (12a) tells us that if the attitude ascription (7) is true, then Mary asserts / believes some claim that is necessarily equivalent to the claim that John speaks Spanish; while (12b) assures us that if (7) is not true, then there is something necessarily equivalent to the claim that John speaks Spanish that Mary does not assert / believe. However, the theorems are still not informative enough, since (12a) doesn't tell us which of the many claims necessarily equivalent to that claim that John speaks Spanish Mary asserts / believes, when (7) is true; and (12b) does not tell us which of these claims she fails to assert / believe, when (7) is not true.[10]

Because the information carried by theorems of the sort (12a,b) is still incomplete, we need to see if the theory can be made to provide further information that could be used to determine whether or not the original statements like (9) that it makes about the truth conditions of attitude ascriptions are correct. One type of information standardly provided by a semantic theory is information about which object-language sentences follow from which. In the case of propositional attitude ascriptions, the information we want concerns cases in which the truth of one attitude ascription, together perhaps with the truth of other supplementary premises, does, or does not, guarantee the truth of other attitude ascriptions. Indeed, it was precisely in order to deliver certain results of this kind that Larson and Ludlow formulated the theory in the way that they did.

Foremost in their minds was the desire to provide a semantics that would block inferences like the one from (13a,b) to (13c).

13a. Max believes Judy Garland sang "Somewhere Over the Rainbow."

[10] There is a further problem that arises at this point that I will simply note, without pursuing. When we convert Larson's and Ludlow's theory into one that provides truth conditions relative to alternative worlds, we get the result that 'Mary believes that John speaks Spanish' is true w.r.t. a world w iff in w Mary bears the belief relation to the interpreted logical form (1b). Intuitively, however, it seems that the claim that Mary believes that John speaks Spanish could be true w.r.t. w even though in w the words Mary uses to formulate that belief are not the same as the words she uses in the actual world. In fact in w Mary may not be familiar with any of the words referred to in (1b). Does this mean that in w she doesn't bear the belief relation to (1b)? If so, we get the counterintuitive result that in w Mary doesn't believe that John speaks Spanish, simply because in w she doesn't use all the words 'John', 'speak' and 'Spanish'. If not, then we need to be told how one can bear the belief relation to an interpreted logical form that contains words with which one is not familiar.

b. Judy Garland = Frances Gumm.

c. Max believes Frances Gumm sang "Somewhere over the Rainbow."

Their account of why this inference is blocked is as follows: although the truth of (13b) guarantees that the semantic values of the names *Judy Garland* and *Frances Gumm* are the same, and although this semantic value—i.e. the woman named by both of these expressions—is one constituent of the interpreted logical forms that are supposed to be the objects of the beliefs reported by (13a) and (13c), the interpreted logical forms are different in the two cases because one of them contains the name *Judy Garland* while the other contains the name *Frances Gumm.* Since (13a) reports that Max stands in the belief relation to one theoretical entity, and (13c) reports that Max stands in the belief relation to a different theoretical entity, Larson and Ludlow conclude that the truth of (13a,b) does not guarantee the truth of (13c).

However, there is a problem here. The semantic theory tells us that sentence (13a) is true iff Mary believes a certain interpreted logical form and that (13c) is true iff Mary believes a slightly different interpreted logical form. But nothing in the semantic theory, or in any theory provided by Larson and Ludlow, tells us whether it is possible for an agent to bear the belief relation to one of these interpreted logical forms without bearing the belief relation to the other interpreted logical form. Moreover, since the whole idea of believing one of these specially constructed objects is theoretical, we have no intuitions to consult to settle the matter. In principle, one could combine the Larson and Ludlow semantic account with any one of many different theses about the belief relationship between agents and interpreted logical forms.

Two such theses are given in (14).

14a. The belief relation is such that whenever two interpreted logical forms differ in any way whatsoever it is possible some agent to believe either one without believing the other.

b. The belief relation is such that whenever two interpreted logical forms agree entirely in their non-lexical constituents and differ only in the lexical material they contain, it is impossible to believe one without believing the other.

If one were to combine our extended version of the Larson and Ludlow semantics with (14a) one would get the conclusion that it is possible for (13a,b) to be true while (13c) is false. On the other hand, if one were to combine the semantics with (14b), one would get the conclusion that it is not possible for (13a,b) to be true while (13c) is false. Since Larson and Ludlow do not explicitly adopt either (14a), (14b), or any other comparable principle, their theory makes **no** predictions about when substitution in attitude ascriptions is necessarily truth preserving.

Since their theory needs to make such predictions in order to be properly testable, this is unacceptable. It is also, I think, not their real intent. Although Larson and Ludlow do not explicitly formulate the needed principle, they often speak as if they take it for granted that any difference in interpreted logical forms arising from substitution of one expression for another in attitude ascriptions gives rise to a possible divergence in truth values.[11] If this is their intent, then perhaps we should modify their explicit theory by conjoining it with (14a). This has the virtue of giving the theory needed predictive content. Unfortunately, however, many of the predictions turn out to be false.

For example, consider the pair (15a,b).

15a. Mary believes that John speaks Spanish.

b. María cree que Juan habla español.

A Larson and Ludlow style semantic theory for English will tell us that (15a) is true iff Mary believes the interpreted logical form (1b). A similar semantic theory for Spanish will tell us that (15b) is true iff Mary believes the interpreted logical form (16).

[11] For example, on page 1002, Ludlow and Larson say the following: "The assumption that ILFs [interpreted logical forms] are composed of linguistic forms and extra-linguistic objects yields straightforward individuation criteria for ILFs. Two ILFs will be distinct whenever they contain distinct forms or distinct objects. This in turn yields straightforward criteria for distinguishing attitude reports. Two attitude reports will be logically nonequivalent whenever their complement clauses are associated with ILFs that differ either in form or content." Note, in saying that attitude ascriptions containing different linguistic forms are logically nonequivalent they seem to be saying that it is genuinely possible for one to be true when the other is not. They might, of course, argue that logical possibility is one thing and metaphysical possibility is another. However, it is not clear whether or how they would draw that distinction. Moreover, the quoted comment is followed by a long discussion of instances in which they take the smallest differences in linguistic form in two complement clauses to lead to a difference in truth value in attitude reports.

16.

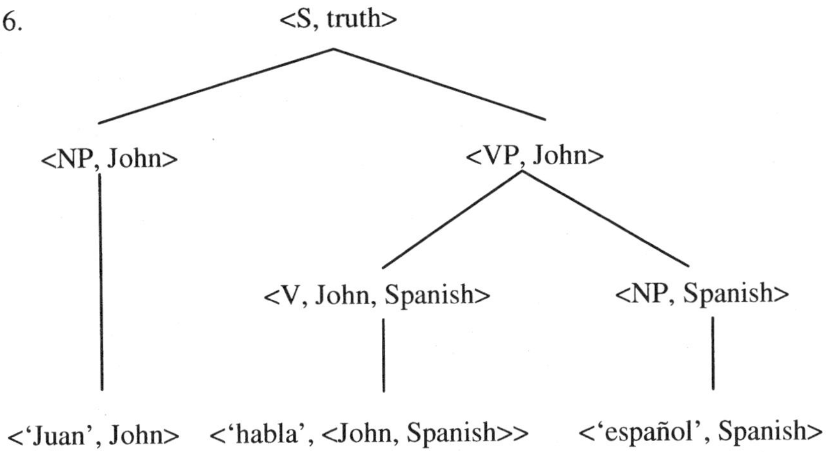

Since (1b) differs lexically from (16), the semantic theories for the two languages together with (14a) predict that it is possible for (15a) and (15b) to differ in truth value. The same predication is made for every pair of attitude ascriptions in different languages. This prediction is, I take it, false.[12]

Additional instances of what seems to be the same problem are given by the following pairs.

17a. There is a professor x in my department such that Mary believes that x speaks Spanish.

b. There is a professor y in my department such that Mary believes that y speaks Spanish.

18a. There is a man and there is a woman such that Mary believes that the former loves the latter.

b. There is a woman and there is a man such that Mary believes the latter loves the former.

[12] In the final section of their paper (section 7), Larson and Ludlow acknowledge the problems posed by examples like (15a,b), as well as other instances in which belief ascriptions with linguistically different complements would normally be counted as "saying the same thing", or having the same content. However, their discussion of this is inconclusive.

19a. Mary believes that you are interesting to talk to. (said to me)

b. Mary believes that I am interesting to talk to. (said by me)

20a. Mary believes it will rain in Madrid tomorrow. (said yesterday)

b. Yesterday, Mary believed it would rain in Madrid today. (said today)

21a. Mary believes that Peking is one of the world's largest cities, but she doesn't know its name.

b. Mary believes that Beijing is one of the world's largest cities, but she doesn't know its name.

In each case, the interpreted logical form which is the alleged object of the (a)-ascription differs linguistically from the interpreted logical form that is the alleged object of the (b)-ascription. Hence the Larson and Ludlow semantics, augmented with (14a), predicts that the two ascriptions may differ in truth value. This seems wrong.

Similar problems are posed by (22) - (24).

22a. Mary believes that I / you speak Spanish (said by me)

b. There is someone such that Mary believes that that person speaks Spanish.

23a. Mary believes that thing is heavy. (said demonstrating an anvil)

b. There is something such that Mary believes that it is heavy.

24a. Mary believes that Kripke wrote *Naming and Necessity*; moreover Kripke still lives in Princeton.

b. There is someone who still lives in Princeton such that Mary believes he wrote *Naming and Necessity*.

In each case, the (b) sentence is a consequence of the (a) sentence. Thus, if the (a) sentence is true, then the (b) sentence must be true. Since the enriched theory consisting of the Larson and Ludlow's semantics plus principle (14a) conflicts with this conclusion, the enriched theory is false.[13]

The lesson here goes far beyond Larson and Ludlow, and cannot be stressed enough. There is a class of celebrated cases, on which philosophers and semanticists have concentrated, in which substitution of coreferential names, indexicals, or variables (relative to contexts and assignments) in attitude ascriptions appears to change truth value. However, in addition to this there is another large class of uncelebrated cases in which such substitution seems clearly to preserve truth value. Both sorts of cases deserve to be taken seriously. The first class of cases might tempt one to adopt the working hypothesis that objects of the attitudes contain linguistic as well as nonlinguistic elements. If one does this, however, one must also account for the second class of cases by formulating clear principles about precisely when and how the linguistic elements incorporated into the objects of the attitudes should be ignored, because they make no difference to the truth value of attitude ascriptions, and when they should not be ignored, because they do make a difference. Moreover, there is no guarantee in advance of a systematic investigation of this kind that linguistic elements will prove to be needed in the objects of the attitudes at all.

3. EXTENDING THE PROBLEM OF EXTRACTING TESTABLE EMPIRICAL CONTENT TO OTHER SEMANTIC THEORIES

In addressing the problem as applied to the theory of Larson and Ludlow, I proceeded in three steps.

(i) I began by noting that the truth conditions assigned to propositional attitude ascriptions by T-sentences like (9) are themselves theoretical claims about which ordinary linguistic intuitions provide no guidance. These T-sentences can be evaluated only by combining them with further semantic principles in a way that generates more obvious and empirically testable statements of truth conditions.

[13] All of these problems are compounded by the observation in footnote 10.

(ii) I extended the theory so as to produce testable results like (11a,b) and (12a,b).

(iii) I supplemented these results with predictions about when, according to the theory, the truth of one attitude ascription, together perhaps with the truth of supplementary premises, guarantees the truth of other attitude ascriptions.

The main problem for Larson and Ludlow came at step (iii). According to them, belief ascriptions report relations between agents and interpreted logical forms. Thus, whenever the interpreted logical forms of complement sentences of two belief ascriptions differ in any way, the ascriptions are taken to report relations to different objects of belief. This raises a question. When, if ever, does the fact that an agent bears the belief relation to one interpreted logical form guarantee that the agent bears the belief relation to a different interpreted logical form. We saw that, without an answer to this question, the theory makes no predictions about when one belief ascription follows from another, and so remains too empirically underspecified to be of significant interest. We also saw that if we were to supplement their explicit theory with a principle characterizing the belief relation as one that can, in principle, always hold between an agent and one interpreted logical form while failing to hold between that agent and another interpreted logical form, no matter how slightly the second differs from the first, then their theory must be rejected for making false predictions about the relationship between the truth conditions of different belief ascriptions.

Nor will it help to utter the superficially pleasing words, "an agent believes the interpreted logical form expressed by the complement sentence of a belief ascription uttered in a given context iff the agent bears an analogous relation to a *relevantly similar* interpreted logical form," and leave it at that. Unless one can offer an independent, substantive account of what the analogous relation is and what it is for one interpreted logical form to be relevantly similar to another, this is just a cheat that empties the theory of empirical content, without acknowledging that this is what is going on. In the end, there is no avoiding the fact that we cannot accept any semantic theory of propositional attitude ascriptions in the absence of clear and definite predictions of the kind mentioned in step (iii).

In order to demonstrate the generality of this point, I will close by applying the logic of steps (i), (ii), and (iii) to two other analyses of propositional attitude ascriptions drawn from richer semantic frameworks. Both of these analyses accept the relational principle, (R).

R. A propositional attitude ascription ***x v's that S*** is true (relative to a context C, world w of evaluation, and assignment A of i to 'x') iff in w, i bears the relation R_v (expressed by v in C) to the proposition expressed by S (relative to C and A).

The two analyses differ in what they take propositions to be.

According to the strict possible-worlds analysis, propositions are sets of possible worlds, and a semantic theory takes the form of a theory of truth relative to a context and a world.[14] The proposition expressed by a sentence S relative to a context C and assignment A is identified with the set of worlds w in which S is true relative to C and A, and the truth conditions for belief ascriptions are given by statements the form (25).

25. A sentence ***x believes that S*** is true relative to a context C, world of evaluation w, and assignment A of i to 'x' iff in w, i believes (i.e. bears the belief relation to) the set of possible worlds w' such that S is true relative to C, w' and A.

Statements of this form are what correspond in the possible worlds account to theorems of the kind illustrated by (9) in Larson's and Ludlow's theory.

In contrast to the possible worlds analysis, the structured propositions approach takes propositions to be structured complexes the constituents of which are objects, properties and propositional functions. On this approach a semantic theory is a two-stage affair. First there is a recursive assignment of structured propositions to sentences relative to contexts. Second, there is a theory of truth relative to possible worlds for propositions. An example of a simple theory of this kind is given by the conjunction of (26) - (28).

26a. The proposition expressed by an atomic formula $Pt_1...t_n$ relative to a context C and assignment A is < P*, < $o_1...o_n$ > >, where P* is the property expressed by P, and o_i is the referent of t_i relative to C and A. ($t_1...t_n$ are names, variables or simple indexicals.)

b. The proposition expressed by a formula $[\lambda v\ S]\ t$ relative to C and A is < g, <o> >, where o is the referent of t relative to C and A, and g is the function from individuals o' to propositions expressed

[14] In identifying propositions with sets of worlds, rather than functions from worlds to truth values, I am assuming (in order to simplify the discussion) that for every proposition p, the set of possible worlds in which p is definitely true and the set of possible worlds in which it is definitely false jointly exhaust all the possible worlds.

by S relative to C and an assignment A' that differs from A at most in assigning o' as the referent of v.

c. The propositions expressed by ***~S*** and ***S&R*** relative to C and A are < NEG, Prop S > and < CONJ, < Prop S, Prop R > > respectively, where Prop S and Prop R are the propositions expressed by S and R relative to C and A, NEG is the property of being a proposition that is not true and CONJ is the property of being a pair of true propositions.

d. The proposition expressed by ***∃v S*** relative to C and A is < SOME, g >, where SOME is the property of being a propositional function that assigns a true proposition as value to some object o as argument, and g is as in (b).

e. The proposition expressed by ***t believes that S*** relative to C and A is < B, < o, Prop S> >, where B is the belief relation, o is the referent of t relative to C and A, and Prop S is the proposition expressed by S relative to C and A.

f. The proposition expressed by a sentence (with no free variables) relative t to C is the proposition it expresses relative to C and every assignment A.

27a. A proposition is < P*, < $o_1 \ldots o_n$ > > is true with respect to a world w iff < $o_1 \ldots o_n$ > is a member of the extension of P* with respect to w.

b. A proposition < g, <o> > is true with respect to w iff g(o) is true with respect to w.

c. A proposition < NEG, Prop S> is true with respect to w iff Prop S is not true with respect to w. A proposition < CONJ, <Prop S, Prop R > > is true with respect to w iff <Prop S, Prop R > is a pair of propositions that are true with respect to w.

d. A proposition < SOME, g> is true with respect to w iff g(o) is true with respect to w, for some object o in w.

e. A proposition < B, < o, Prop S> > is true with respect to w iff in w, o believes Prop S.

28. Here we specify for each name and indexical, what it refers to relative to an arbitrary context, for each predicate what property it expresses, and for each such property it's extension at an arbitrary world.

Together, (26), (27) and (28) generate statements of the truth conditions of sentences, including propositional attitude ascriptions, relative to contexts and worlds.[15]

We are now ready to apply the problem of extracting testable empirical content to our two new theories of the semantics of attitude ascriptions. Step (i) in this process is just as it was for Larson and Ludlow.

(i) The truth conditions of propositional attitude ascriptions are given by statements like (29a), in the case of the possible worlds semantic analysis, and (29b) in the case of the structured propositions semantic analysis. Like the canonical statement (9) of truth conditions derived by Larson and Ludlow, (29a) and (29b) are theoretical claims about which ordinary linguistic intuition provides no guidance. These statements can be evaluated only by combining them with further principles in a way that generates empirically testable claims.

29a. The sentence *John believes that someone loves Mary* is true relative to a context C and world w iff in w John believes (bears

[15] For further details and discussion of semantic theories of structured propositions see S. Soames, "Direct Reference, Propositional Attitudes, and Semantic Content," *Philosophical Topics*, Vol. 15, No. 1, 1987, pp. 47-87, reprinted in *Propositions and Attitudes*, edited by N. Salmon and S. Soames, (Oxford: Oxford University Press, 1988); S. Soames "Direct Reference and Propositional Attitudes," in *Themes from Kaplan*, edited by J. Almog, J. Perry, and H. Wettstein, (New York and Oxford: Oxford University Press, 1988), S. Soames, "Semantics and Semantic Competence, *Philosophical Perspectives*, 3, 1989, pp. 575 – 596," and N. Salmon, *Frege's Puzzle*, (Cambridge, MA.: M.I.T. Press, 1986).

the belief relation to) the set of possible worlds in which someone loves Mary. (Strict Possible Worlds Analysis)

b. The sentence *John believes that someone loves Mary* is true relative to a context C and world w iff in w John believes (bears the belief relation to) the proposition that is the ordered pair the first coordinate of which is the property SOME, of being a propositional function that assigns a true proposition as value to some object o as argument, and the second coordinate of which is the function g that assigns to any object o the proposition that is the ordered pair the first coordinate of which is the relation of loving and the second coordinate of which is the ordered pair the first coordinate of which is o and the second coordinate of which is Mary. (Structured Propositions Analysis)

Next we move to step (ii).

(ii) Just as the theory of Larson and Ludlow can be extended to generate testable empirical claims like (12a,b), so the possible worlds analysis and the structured propositions approach will generate testable claims like (30a,b).[16]

30a. If the sentence *John believes that someone loves Mary* is true relative to a context C and world w, then there is something p that is true with respect to an arbitrary world w* iff someone loves Mary in w*, and in w John believes p.

b. If the sentence *John believes that someone loves Mary* is not true relative to a context C and world w, then there is something p that is true relative to an arbitrary world w* iff someone loves Mary in w*, and in w John does not believe p.

Finally, although predictions like (30a,b) made by these theories are correct, they are too meager to provide conclusive confirmation of the theories.[17] To provide a further empirical test of the theories, we need to

[16] In the case of the possible worlds approach we need to specify that a proposition (i.e. set of worlds) p is true w.r.t. an arbitrary world w iff w is a member of p.

[17] Of course the possible worlds semantic analysis makes a stronger prediction as well. It predicts that the sentence *John believes that someone loves Mary* is true relative to a

move to step (iii), where we examine predictions about when, according to the theories, the truth of one attitude ascription, together perhaps with the truth of supplementary premises, guarantees the truth of other attitude ascriptions. This gives us the following results.

(iii) The possible worlds semantic analysis of attitude ascriptions must be rejected because it wrongly predicts: (a) that when S and S' are necessarily equivalent (with respect to C and A) ***John believes that S*** and ***John believes that S'*** must agree in truth value (with respect to C and A), and (b) that since *believe* distributes over conjunction, whenever S' is a necessary consequence of S (w.r.t. C and A), *John believes that S'* is a necessary consequence of ***John believes that S*** (w.r.t. C, A). The structured propositions approach avoids these difficulties, as well as those posed for Larson and Ludlow by (15) – (24).

This result highlights what I take to be the promise of the approach to the semantics of propositional attitude ascriptions taken by the structured propositions analysis. In order for that promise to be realized two further problems must be solved. First, additional principles must be added to the theory in order to make further correct predictions about the truth conditions of various attitude ascriptions. For example, a principled account of the belief relation needs to be given which has the consequence that anyone who believes a conjunction thereby believes both conjuncts. Given such a principle, the structured proposition analysis will correctly predict results such as (31).

31. For any sentences S and R, and for any context C and world w, if ***x believes that S and R*** is true relative to C, w and an assignment A of an individual as value of 'x', then ***x believes that S*** and ***x believes that R*** are also true relative to C, w, and A.

Additional principles covering constructions other than conjunction, and attitudes other than belief, are needed to generate further results of this type. The second problem to be solved involves finding a way either to accommodate, or to explain away, apparent cases in which substitution of one coreferential name or indexical for another in the complement of an

context C and a world w iff in w John believes something that is true with respect to an arbitrary world w* iff someone loves Mary in w*. However this prediction is false, since, contrary to the possible worlds approach, a person can believe a proposition p that is true in all and only the worlds in which q is true, without believing q.

attitude ascription changes truth value.[18] Although solving these problems is no easy task, I believe it can be accomplished. However the argument for this is something that must be left for another time.[19]

[18] For an attempt to deal with apparent instances of substitution failure in propositional attitude ascriptions see my treatment of the issue in *Beyond Rigidity: The Unfinished Semantic Agenda of Naming and Necessity*, (New York: Oxford University Press, 2002).

[19] The material in this paper extends and develops a point sketched in the first few pages of chapter 7 of *Beyond Rigidity*. I would like to thank the commentators on my paper and the participants at the ICCS-01 conference for useful comments. Timothy Williamson, in particular, made a helpful observation concerning examples (11) and (12).

Chapter 3

TRUTH AND BORDERLINE CASES

Timothy Williamson
University of Oxford

1. INTRODUCTION

According to the principle of bivalence, truth and falsity are jointly exhaustive and mutually exclusive options for a statement. It is either true or false, and not both, even in a borderline case. That highly controversial claim is central to the epistemic theory of vagueness, which holds that borderline cases are distinguished by a special kind of obstacle to knowing the truth-value of the statement. But this paper is not a defence of the epistemic theory. If bivalence holds, it presumably does so as a consequence of what truth and falsity separately are. One may therefore expect bivalence to be derivable from a combination of some principles characterizing truth and other principles characterizing falsity. Indeed, such derivations are easily found. Their form will of course depend on the initial characterizations of truth and falsity, and not all such characterizations will permit bivalence to be derived. This paper focusses on the relation between its derivability and some principles about truth and falsity. Borderline cases for vague expressions are primary examples of an urgent challenge to bivalence.

A key variable in the relation is obviously the choice of a logic. The background logic here is classical. That choice does not automatically prejudge the issue in favour of bivalence, for the latter can fail even in a classical context on some non-standard accounts of truth and falsity; traditional supervaluationist semantics provides the best-known instance.

K. Korta and J. M. Larrazabal (eds.), Truth, Rationality, Cognition, and Music, 45-66.

My strategy is to start not with semantic theories designed to have some particular result for vague languages but with principles which seem natural from the standpoint of the theory of truth and falsity. I then explore their implications for bivalence. More specifically, simple considerations about the semantics of indexicals motivate replacing the usual disquotational characterization of truth by one which makes explicit allowance for contextual variation in what is said by a sentence. The apparatus I use for that purpose *appears* to make room for failures of bivalence in borderline cases. For the argument for bivalence requires a principle of uniformity in what a given sentence says in a given context, and borderline cases may appear to motivate a denial of such uniformity. I will argue that the appearance is illusory.

2. SAYING AND DISQUOTATION

A formal framework is needed in which to conduct the investigation. Since the concern is with vagueness in language, I will treat sentences as the bearers of truth and falsity. That is consistent with the idea that the truth-value of a sentence is determined by the truth-value of a proposition which it expresses. Of course, a sentence may express different propositions with respect to different contexts of utterance, if it contains indexicals or demonstratives. Thus the truth-value of a sentence is also relative to a context of utterance. The expression **Say(s,c,P)** will mean: the sentence s as uttered in the context c says (or is used to say) that P (where 'P' may be replaced by a declarative sentence); in terms of propositions, s in c expresses the proposition that P. **True(s,c)** (respectively, **False(s,c)**) means: s is true (respectively, false) in c. One could easily extend the discussion from vagueness in language to vagueness in non-linguistic thought by letting **Say** mean the more general notion of expressing and s range over thought types as well as sentences.

I will take for granted the principles of classical logic, appealing without comment to substitution instances of theorems of the classical propositional calculus and its extensions for the relevant types of quantifier, and use the rule of *modus ponens* similarly. Although these logical assumptions are of course not uncontroversial in the case of vagueness, justifying them is not my current concern. They are in any case not exclusive to a particular theory of vagueness; they are common to epistemicism, supervaluationism and some other views. Moreover, as it happens, many — but not all— of the arguments will use only uncontroversial fragments of classical logic.

Within this framework, the natural principle about truth is that if a sentence says that something is so, then it is true if and only if that thing is so. More precisely:

(T) $\forall s \forall c \forall P[\text{Say}(s,c,P) \supset [\text{True}(s,c) \equiv P]]$

(T) is not itself a disquotational principle about truth, for it contains no occurrence of **P** in quotation marks (explicit or implicit). To recover a disquotational biconditional of the form **True(«P»,c) ≡ P** from (T), one would need a corresponding auxiliary premise of the form **Say(«P»,c,P)**. Such a premise holds if the sentence which replaces **P** contains no context-dependent elements. For example, if '$E = mc^2$' is context-independent, then in any context it says that $E = mc^2$, and so is true if and only if $E = mc^2$. Even if **P** is replaced by a context-dependent sentence, the auxiliary premise **Say(«P»,c,P)** still holds if **c** refers to the context in which that premise is being asserted. For example, as uttered by me 'I am Timothy Williamson' says that I am Timothy Williamson, and so is true if and only if I am Timothy Williamson. But 'I am Timothy Williamson' as uttered by you does not say that I am Timothy Williamson, for 'I' as uttered by you does not refer to me. The sentence as uttered by you is not true. The disquotational biconditional holds only under restricted circumstances. But (T) does better; it handles context-dependence with ease. Since 'I am Timothy Williamson' as uttered by you does not say that I am Timothy Williamson, the corresponding instance of (T) is vacuously true. That sentence as uttered by you does say that you are Timothy Williamson, and it is true as so uttered if and only if you are Timothy Williamson. Thus (T) is more basic than the disquotational biconditional; it explains both the successes and the failures of the latter.

A further advantage of (T) over the disquotational biconditional is that the latter but not the former must be revised to meet semantic paradoxes such as the Liar. For example, in some context c one can construct a self-referential sentence s to be **~True(s,c)**. The corresponding disquotational biconditional is **True(«~True(s,c)»,c) ≡ ~True(s,c)**; since **s = «~True(s,c)»** holds, one can deduce **True(s,c) ≡ ~True(s,c)**, which is logically false. Thus not every instance of **True(«P»,c) ≡ P** holds. But this argument does not show that (T) is invalid; it merely falsifies the antecedent **Say(«~True(s,c)»,c,~True(s,c))**. Not even the sentence 'This sentence is not true' succeeds in saying of itself that it is not true (for further discussion see Williamson 1994: 197 and 1998).

The variable **P** takes sentence position in (T). In particular, it flanks the biconditional ≡. Correspondingly, the third argument place in **Say(s,c,P)** is for a sentence, not a name (not even a name of a sentence); with respect to that argument place, **Say** works like an operator rather than a (first-level) predicate. Thus (T) involves quantification into sentence position. Such quantification might be interpreted substitutionally. The sentences substituted for **P** would be declarative sentences of the language in which the theorist is working (English, say). This would not restrict the instances of (T) which hold non-vacuously to those in which the sentence to which **s** refers is itself a sentence of English, for we can express in English what a sentence of some other language says. For example, the Serbian sentence 'Jedan sto je u sobi' as uttered in a suitable context says that one table is in the room, and so is true if and only if one table is in the room. However, if in some context a sentence s says something which cannot be expressed in English, at least in the context in which the theorist is working, then no instance of (T) holds non-vacuously for that case. (T) would still be true, but it would not be as informative about such cases as we should like it to be. We might therefore consider a non-substitutional reading of the quantifier **∀P**, on which **∀P φ(P)** might be false even if **φ(P)** was true whenever **P** is replaced by a sentence of the language in which the theorist is working. For example, given fixed assignments to the variables **s** and **c**, **∀P ~Say(s,c,P)** might be false even though every corresponding substitution instance of the form **~Say(s,c,P)** was true. Such a non-substitutional interpretation would *not* automatically be objectual, for example over all propositions conceived as objects of a special kind. Objectual quantification is quantification whose semantics is given by non-substitutional quantification *into name position* in the metalanguage; both 'proposition' and 'object' are nouns, not sentences. One should not dismiss without argument the possibility that non-substitutional quantification into sentence position is irreducible because its semantics can be given faithfully only by non-substitutional quantification into sentence position in the metalanguage too. It is not obviously wrong to suppose that one can understand such non-objectual non-substitutional quantification even if it cannot be expressed unequivocally in English; that might be a defect of English. We shall not attempt to decide between the substitutional and non-substitutional options here. Both are consistent with the arguments below. Even if the substitutional interpretation is not the intended one, it still provides a consistency proof of a standard theory of quantification into sentence position. The use made below of such quantification is quite undemanding, for sentence variables are not replaced by sentences that themselves contain either quantification into sentence position or one of the expressions **True**, **False** or **Say**.

Corresponding to (T), the natural principle about falsity is that if a sentence says that something is so, then it is false if and only if that thing is not so. More precisely:

(F) $\forall s \forall c \forall P[\mathbf{Say}(s,c,P) \supset [\mathbf{False}(s,c) \equiv \sim P]]$

For example, in a context in which the Serbian sentence 'Jedan sto je u sobi' says that one table is in the room, it is false if and only if it is not the case that one table is in the room. One needs the same auxiliary premise **Say(«P»,c,P)** to infer the disquotational biconditional **False(«P»,c) ≡ ~P** from (F) as one does to infer the corresponding disquotational biconditional **True(«P»,c) ≡ P** from (T). Much more will be said about both (T) and (F) below.

3. BIVALENCE

How should the principle of bivalence be formulated within this framework? The principle should not imply that non-declarative sentences are true or false, for presumably they are not intended to say that something is the case. For the same reason, the principle does not imply that a declarative sentence is true or false if it does not say that something is the case. For example, if a language teacher writes the sentence 'That belongs to her' on the board to illustrate a point of grammar, without attempting to supply a reference for the demonstratives, the principle of bivalence does not require the sentence to be false. Thus one can reasonably build into the antecedent condition that the sentence says that something is the case, just as in (T) and (F). The usual formulation then says, on that condition, that the sentence is either true or false, where the disjunction is understood as inclusive. Since classical logic is being assumed, that is equivalent to the claim that the sentence is false if it is not true:

(WB) $\forall s \forall c \forall P[\mathbf{Say}(s,c,P) \supset [\sim\mathbf{True}(s,c) \supset \mathbf{False}(s,c)]]$

Call (WB) the principle of *Weak Bivalence*. It is weak because, by itself, it is consistent with the supposition that (for each fixed context) **True** and **False** stand for exactly the same property, one possessed by every sentence saying that something is so, even if there are many such sentences. In that case **True** and **False** would stand for one and the same truth-value, not two, and the term 'bivalence' would be a misnomer. One can strengthen weak bivalence by adding its converse, the principle that truth and falsity are mutually exclusive for a sentence that says that something is so:

(ME) $\forall s \forall c \forall P[\mathbf{Say}(s,c,P) \supset [\mathbf{False}(s,c) \supset \sim\mathbf{True}(s,c)]]$

Suppose that a sentence is true only if it says that something is so; then one can drop the antecedent **Say(s,c,P)** and (ME) implies the simpler principle that truth and falsity are unconditionally mutually exclusive. (WB) and (ME) can be combined into a single principle of *Strong Bivalence*:

(SB) $\forall s \forall c \forall P[\mathbf{Say}(s,c,P) \supset [\sim \mathbf{True}(s,c) \equiv \mathbf{False}(s,c)]]$

In classical logic, the combination of the biconditional and negation can be reformulated as an exclusive disjunction: a sentence that says that something is so is either true or false and not both. (SB) stands to (WB) as exclusive stands to inclusive disjunction. For any context in which at least one sentence says that something is so, (SB) requires **True** and **False** to stand for distinct properties, and the term 'bivalence' is appropriate. Contrary to the usual practice, I will concentrate on the strong rather than weak form of bivalence. Of course, (SB) is equivalent to (WB) given the almost universally accepted principle (ME). Nevertheless, (SB) does better than (WB) in making some logical and philosophical connections salient (Williamson (1994, 1995 and elsewhere) uses 'bivalence' for weak bivalence; the idea of working with strong bivalence was proposed in an earlier version of Andjelković (1999)). Dialetheists think that paradoxical sentences such as the Liar are both true and false; on some readings they accept (WB) but reject (SB) and (ME). I discuss the status of (SB), (WB) and (ME) below.

The logical relations between (T), (F) and (SB) are rather simple: any two of them entail the third. To check that, note that their main biconditionals are (classically equivalent to) $\sim \mathbf{True}(s,c) \equiv \sim \mathbf{P}$, $\mathbf{False}(s,c) \equiv \sim \mathbf{P}$ and $\sim \mathbf{True}(s,c) \equiv \mathbf{False}(s,c)$ respectively and that the biconditional behaves symmetrically and transitively. Thus, given (T), (F) is equivalent to (SB); given (F), (T) is equivalent to (SB); given (SB), (T) is equivalent to (F).

For future reference, it is convenient to label the two directions of (T) and (F) separately:

(T→) $\forall s \forall c \forall P[\mathbf{Say}(s,c,P) \supset [\mathbf{True}(s,c) \supset \mathbf{P}]]$

(T←) $\forall s \forall c \forall P[\mathbf{Say}(s,c,P) \supset [\mathbf{P} \supset \mathbf{True}(s,c)]]$

(F→) $\forall s \forall c \forall P[\mathbf{Say}(s,c,P) \supset [\mathbf{False}(s,c) \supset \sim \mathbf{P}]]$

(F←) **∀s∀c∀P[Say(s,c,P) ⊃ [~P ⊃ False(s,c)]]**

One can easily check that these logical relations hold amongst the unidirectional principles:

(T←) and (F←) entail (WB);

(T→) and (F→) entail (ME);

(F←) and (ME) entail (T→);

(F→) and (WB) entail (T←);

(T←) and (ME) entail (F→);

(T→) and (WB) entail (F←).

By considering unintended interpretations of **True** and **False**, one can check that no further entailments obtain amongst the six halves of (T), (F) and (SB) beyond those implicit in the foregoing list, if **True** and **False** are treated as primitive predicates, not as logical constants (Andjelkovic´ and Williamson 2001).

Epistemicists about vagueness typically argue from (T) and (F) to (SB), or more specifically from (T←) and (F←) to (WB) (Williamson 1994: 188-189 and 1995). Someone could consistently accept both (T) and classical logic while still rejecting (F), and therefore (SB), and therefore epistemicism (Andjelković (1999) points out that such a combination is possible; Williamson (1999) responds). Thus the notion of falsity should play a significant role in the discussion of vagueness; principles such as (F) need explicit attention. Unfortunately, the literature has neglected (F) and cognate principles about falsity.

4. FALSITY AND NEGATION

Can the notion of falsity be defined in terms of the notion of truth? If so, one might be able to use the definition to reduce (F) to (T), and therefore to derive (SB) from (T). In that case, the neglect of the notion of falsity would be more apparent than real, since the notion would have been treated implicitly in the treatment of the notion of truth. Presumably such a definition should not use auxiliary conceptual resources so strong that they would suffice by themselves for a direct definition of falsity without the detour through truth.

The simplest attempted definition of falsity is as non-truth: **False(s,c)** $=_{def}$ **~True(s,c)**. Under this definition, (SB) becomes a trivial logical truth, and (F) reduces to (T). However, the definition has very implausible consequences. A sentence which does not say that something is so counts as false simply because it is not true. Indeed, mountains and lakes count as false for the same reason. One could revise the definition by defining falsity as the conjunction of non-truth with saying that something is so: **False(s,c)** $=_{def}$ **~True(s,c) & ∃P Say(s,c,P)**. According to epistemicism, such a definition would give at least extensionally correct results. But it would not be acceptable to those who reject (SB); a less controversial definition would be preferable. Moreover, it employs quantification into sentence position; given such quantification, one might as well define falsity directly, rather than making an unnecessary detour through the notion of truth (see below).

Falsity is often defined as the truth of the negation (see below for what counts as a negation). Such a definition of falsity may be understood as implying that an item is false only if it has a true negation, and therefore only if it has a negation, so that it is not false if it has no negation at all. Since lakes and mountains have no negations, they are not false. If a sentence s fails to say that something is so, then either s lacks a negation or its negation also fails to say that something is so, for what the negation of s says would be the contradictory of what s says; either way, s does not count as false.

Let the singular term **Ns** refer to a negation of the sentence to which the term **s** refers. If the latter has several verbally different negations, it does not matter which of them the former refers to. Thus the definition of falsity as the truth of the negation yields this principle:

(FN) $\quad\forall s\forall c[\mathbf{False(s,c)} \equiv \mathbf{True(Ns,c)}]$

The biconditional in (FN) is not conditional on the antecedent **Say(s,c,P)**, for if s fails to say that something is so then, as noted above, Ns also fails to say that something is so; consequently, neither **False(s,c)** nor **True(Ns,c)** holds, and therefore the biconditional holds vacuously.

Obviously, (FN) does not help unless we can independently characterize negation (**N**). The easiest way to do so is by using negation in the metalanguage to say what Ns says:

(N) $\forall s \forall c \forall P[\mathrm{Say}(s,c,P) \equiv \mathrm{Say}(Ns,c,\sim P)]$

Note that, given (N), one cannot automatically assume that a sentence is the negation of its negation. For two applications of (N) yield **Say(s,c,P) ≡ Say(NNs,c,~~P)**; on a fine-grained notion of saying, **Say(s,c,~~P)** is not equivalent to **Say(s,c,P)** (we might still allow that a sentence is a *contradictory* of its negation). On a very fine-grained view, not endorsed here, the negation of a sentence consists of a negation operator and that sentence, the former applied to the latter.

Given (FN) and (N), we can reduce (F) to (T) and thereby derive (SB). For as a special case of (T) we have **Say(Ns,c,~P) ⊃ [True(Ns,c) ≡ ~P]**, which by (N) yields **Say(s,c,P) ⊃ [True(Ns,c) ≡ ~P]**, which by (FN) yields **Say(s,c,P) ⊃ [False(s,c) ≡ ~P]** and therefore (F). Once we have (T) and (F), we can derive (SB).

Even without (FN), one can use (T) and (N) to derive a principle related to (SB) (although without the implication that the sentences takes one of two values) with **True(Ns,c)** in place of **False(s,c)**: in a context in which a sentence says that something is so, either it or its negation is true. Thus the rejection of (F) by itself would not avoid all the controversial consequences of an epistemic theory of vagueness, since one would still be committed to the claim that when a vague sentence in a borderline case says that something is so, either it or its negation is true, and not both, even though we have no idea how to find out which of the two contradictory sentences is true.

Let us return to the original principle (SB). To assess it, one must consider the notion of falsity itself, and therefore (FN). What happens to (FN) if s has no negation? If s is not a candidate for truth or falsity —for example, if s is a lake or mountain— then (FN) holds vacuously. But might not s be a candidate for truth or falsity and still have no negation, because it is in an expressively impoverished language? Creatures might in principle communicate information about their environment to each other in a limited system of signals without negation. Perhaps some animals actually do so. Such signals could be true or false, depending on the relation between the signalled state of the environment and its real state. Suppose, for example, that the animals can signal that there is food over there. Thus, when there is food over there, they can communicate truly that there is food over there.

When there is no food over there, they can communicate falsely that there is food over there. That does not require them to have the capacity to negate a signal; they may be unable to signal that there is no food over there. Of course, questions can be raised about the legitimacy of attributing propositional content to signals in so simple a system. Nevertheless, I see no decisive obstacle to such attributions. Creatures are fallible; the capacity to get things right requires the capacity to get them wrong. If the capacity to negate signals is not strictly necessary for communicating true information, it is also not strictly necessary for communicating false information.

A friend of (FN) might reply that a negation operator N can always in principle be added to the system, so that one can define the falsity of s as the truth of its hypothetical negation Ns. This proposal must be formulated with some care, for the falsity of s in a given situation is equivalent to the truth of its negation in that very situation, not in a counterfactual system in which negation has been introduced into the system. For the presence of negation in the signal system might indirectly affect the aspect of the environment about which information is being communicated, by affecting the behaviour of the animals. Intuitively, the complications of working out what might happen in counterfactual circumstances in which a signal had a negation are quite irrelevant to the truth or falsity of its actual tokens. Nor is the definition of the falsity of s as the *actual* truth of a merely *possible* or abstract negation of s particularly attractive. There is a simpler way.

I propose to treat falsity as a primary notion on a par with truth. If negation has a role to play in the characterization of falsity, it is negation as used in the metalanguage by the theorist, not negation as used in the object-language by the speakers under study. For example, ~ occurs in (F) as the metalanguage negation, whereas N is mentioned in (FN) as the object-language negation. Given (T) and (N), (F) is equivalent to (FN) whenever **Ns** is well-defined, so (F) and (FN) do not compete for correctness, although they may compete for primacy.

Even in a language without negation, vagueness can arise in virtue of borderline cases. In some contexts, a negationless signal may be neither clearly true nor clearly false. A sorites series may be possible for it: for instance, a series of contexts, each indiscriminable from the next, in the first of which it is true and in the last of which it is false. Although one needs logical constants to formulate the sorites reasoning explicitly, one does not need them merely to have difficulty in using the signal correctly in such a series of contexts. For example, the case may be classified differently depending on which end of a sorites series it is approached from. Without the logical constants, one cannot properly reflect on the significance of the phenomena of vagueness, but at least some of the phenomena themselves can occur.

The possibility of vague but negationless signals also suggests that borderline cases should not primarily be conceived as those in which speakers are poised symmetrically between a positive and a negative assertion, since the latter is not always an option. Certainly the pragmatic significance of the absence of a positive assertion does not give it the semantic significance of a negative assertion.

The concepts of truth and falsity are parallel in most respects, but not in all, at least as they are explained by (T) and (F). For (F) uses negation in the metalanguage at a point at which (T) does not use any operator at all. I have not defended the use of negation rather than some other operator to define falsity, although it seems very natural; if (F) is coherent, what does it characterize other than falsity? By substituting other operators in the metalanguage for ~, one can construct a variety of other concepts. For example, if ~ is replaced by $\Box$, read 'necessarily', the result characterizes the concept of necessary truth. If ~ is replaced by Δ~, where Δ is read 'clearly', a concept of clear falsity is characterized, and one should not expect to derive the analogues of (SB) or (WB). Within classical logic, the principle of bivalence stands or falls with the connection between falsity and standard negation.

5. DEFINITIONS OF TRUTH AND FALSITY

One might be tempted to regard (T) and (F) as implicit definitions of **True** and **False** respectively. That would be a mistake. Even granted the correctness of (T) and (F), in one respect they are too weak to constitute definitions; in another respect they are too strong to do so. They guarantee neither uniqueness nor existence.

In what follows we will treat **True** and **False** as so far uninterpreted symbols, but the other primitive terms of the formal language as already interpreted. One can therefore ask how various principles constrain the interpretation of **True** and **False**. (T) and (F) are too weak to define **True** and **False** because they are quite neutral about the application of those terms to a sentence when it says nothing. Presumably a sentence is neither true nor false in a context in which it says nothing, but (T) and (F) impose no such constraint. They are equally consistent with the unappealing supposition that a sentence is both true and false in a context in which it says nothing. Consequently, they do not characterize **True** and **False** uniquely. To be more precise, let (T_1) and (T_2) be the results of subscripting **True** in (T) by '1' and '2' respectively. Then from (T_1) and (T_2) on the assumption **Say(s,c,P)** one can deduce $\mathbf{True_1(s,c) \equiv P}$ and $\mathbf{True_2(s,c) \equiv P}$ and therefore $\mathbf{True_1(s,c) \equiv True_2(s,c)}$, giving:

(T!-) $\forall s \forall c \forall P[\mathbf{Say}(s,c,P) \supset [\mathbf{True}_1(s,c) \equiv \mathbf{True}_2(s,c)]]$.

But one cannot drop the assumption and assert unconditionally:

(T!) $\forall s \forall c[\mathbf{True}_1(s,c) \equiv \mathbf{True}_2(s,c)]$.

Thus truth is not *the* value of **True** which satisfies (T), because there is no such unique value. Similarly, if (F_1) and (F_2) are the results of subscripting **False** in (F) accordingly, they entail:

(F!-) $\forall s \forall c \forall P[\mathbf{Say}(s,c,P) \supset [\mathbf{False}_1(s,c) \equiv \mathbf{False}_2(s,c)]]$.

But one cannot drop the assumption and assert unconditionally:

(F!) $\forall s \forall c[\mathbf{False}_1(s,c) \equiv \mathbf{False}_2(s,c)]$.

Thus falsity is not *the* value of **False** which satisfies (F), because there is no such unique value. We can fill these gaps by explicitly stipulating that a sentence is true or false only if it says something:

(T+) $\forall s \forall c[\mathbf{True}(s,c) \supset \exists P\ \mathbf{Say}(s,c,P)]$

(F+) $\forall s \forall c[\mathbf{False}(s,c) \supset \exists P\ \mathbf{Say}(s,c,P)]$

With these additions, truth and falsity are uniquely characterized, in the sense that (T!) follows from (T_1), (T_2), (T_1+) and (T_2+) (the last two being the results of subscripting **True** in (T+) accordingly), and (F!) follows from (F_1), (F_2), (F_1+) and (F_2+). It is worth noting that (T+) and (F+) rely more heavily on quantification into sentence position than do (T) and (F). For one could drop the universal quantifiers from (T) and (F) and treat the remainders simply as schemata, whereas no such treatment is possible for the existential quantifiers in (T+) and (F+).

A deeper problem is that (T) and (F) are too strong to be mere definitions of **True** and **False**, because they are *creative*. That is, (T) has a new consequence not involving **True** and (F) has a new consequence not involving **False**. More specifically, each of them entails a principle of *uniformity* to the effect that everything said by a given sentence in a given context has the same truth-value:

(U) $\forall s \forall c \forall P \forall Q[[\mathbf{Say}(s,c,P)\ \&\ \mathbf{Say}(s,c,Q)] \supset [P \equiv Q]]$

For the antecedent of (U) yields **True(s,c) ≡ P** and **True(s,c) ≡ Q** by instances of (T), from which **P ≡ Q** follows. Similarly, the antecedent of (U) yields **False(s,c) ≡ ~P** and **False(s,c) ≡ ~Q** by instances of (F), from which **~P ≡ ~Q** follows, and therefore **P ≡ Q** too.

Let L be a language without the predicates **True** and **False** but in which (U) can be formulated, and consider a theory Z in L of which (U) is not a theorem. If one adds **True** to L and (T) to Z, the result is not a conservative extension of Z, because (U) is a theorem of the new theory in the old language without being a theorem of the old theory. For the same reason, if one adds **False** to L and (F) to Z, the result is again not a conservative extension of Z. If (T) and (F) were genuine definitions, their addition would yield conservative extensions of Z. They behave more like theories than definitions.

A question now arises for the epistemicist derivation of the bivalence principles (SB) and (WB). It relies on (T) and (F), which are logically stronger than definitions of **True** and **False**. Thus the opponent of epistemicism might accept classical logic but block the derivation of (SB) and (WB) by rejecting (T) or (F) or both, while insisting that this does not amount to redefining **True** and **False**. In particular, the anti-epistemicist might reject (U), and therefore both (T) and (F), on the grounds that (U) fails in borderline cases for a vague sentence s. Perhaps they will suggest that a vague sentence says many things, corresponding to its different possible sharpenings; in a borderline case, some of these things differ from others in truth-value, contrary to (U).

I will postpone asking whether (U) really does fail in borderline cases, and suppose for the sake of argument that it does. If so, what happens to the bivalence principles (SB) and (WB)? In order to answer that question, one needs characterizations of truth and falsity. (T) and (F) will not do for present purposes because they entail (U), which is being supposed to fail. One must therefore characterize truth and falsity in some alternative way. Within the present framework, the natural idea is to define truth and falsity explicitly by quantifying into sentence position.

A standard proposal is that a sentence is true if something that it says to be so is so. The biconditional corresponding to such a definition is this:

(TDEF1) **∀s∀c[True(s,c) ≡ ∃P[Say(s,c,P) & P]]**

The obvious analogue for falsity of such a definition is that a sentence is false if something that it says to be so is not so. The corresponding biconditional stands to (F) as (TDEF1) stands to (T):

(FDEF1) **∀s∀c[False(s,c) ≡ ∃P[Say(s,c,P) & ~P]]**

Like (T+) and (F+), both (TDEF1) and (FDEF1) make much more essential use of quantification into sentence position than do (T) and (F). Schemata without such quantification are no substitute for (TDEF1) and (FDEF1).

Before continuing with the main line of argument, I pause to consider an objection. By explicitly defining truth, for instance by (TDEF1), are we not liable to incur inconsistency, by Tarski's theorem on the undefinability of truth? Fortunately, we can avoid this danger by not permitting the unrestricted application of the expression **Say** to sentences of the metalanguage as values of the term in its first argument-place (**s**). This still permits **Say** to be applied to sentences of many diverse languages, provided that their content is not of a special metalinguistic kind. Such a restriction requires extensive discussion, but our present concern is not with the semantic paradoxes. Quantification into sentence position does not itself introduce any inconsistency. As noted above, one can give it a consistency proof by using a substitutional interpretation. Indeed, one could give an alternative consistency proof by using a different unintended interpetation in which the semantic value of the sentential variable **P** is simply its truth-value, although that would not permit **Say** to behave in the intended way, as an intensional operator with respect to **P**. One could accommodate some of that intensionality by treating the semantic value of **P** as the set of possible worlds in which it is true, although even that would not capture its full intended meaning, since we take **Say** to behave *hyper*intensionally: a sentence may say that P without saying that Q even though in all possible worlds P if and only if Q. For example, '2+2=4' says that 2+2=4 without saying that 7+5=12, even though all mathematical truths are true in all possible worlds and therefore in the same possible worlds. Given the limited nature of the present use of quantification into sentence position, as noted above, we can consistently treat the semantic values of variables in sentence position as structured or otherwise finely grained propositions, thereby allowing for hyperintensionality. We will not pursue these issues here, but simply repeat that none of the logical framework within which we are operating falls to a semantic paradox.

Since (TDEF1) and (FDEF1) are in effect definitions, they do not entail (U). Therefore, they do not entail (T) or (F). More specifically, one can easily check that (TDEF1) entails (T←), and therefore does not by itself entail (T→) (for if it did, it would entail (T)), and that (FDEF1) entails (F←), and therefore does not by itself entail (F→) (for if it did, it would entail (F)).

We can check that (TDEF1) and (U) together do entail (T→), and therefore (T), by an obvious argument. Similarly, (FDEF1) and (U) together entail (F→), and therefore (F). This shows that (U) is the *strongest* consequence of the conjunction of (T) and (F) to contain neither **True** nor **False**, in the sense that if A contains neither **True** nor **False** and (T) and (F) together entail A then (U) entails A. For (U), (TDEF1) and (FDEF1) together entail (T) and (F); thus if (T) and (F) entail A, then by transitivity (U), (TDEF1) and (FDEF1) entail A; but if one replaces **True** and **False** throughout the deduction by the right-hand sides of (TDEF1) and (FDEF1) respectively, the result trivializes (TDEF1) and (FDEF1) and is still a valid deduction of A from (U), since the premises and conclusion do not contain **True** or **False**. Thus (U) exhausts the non-conservative aspect of (T) and (F).

What are the implications of (TDEF1) and (FDEF1) for bivalence? As already noted, they entail (T←) and (F←) respectively, which together entail (WB), so (TDEF1) and (FDEF1) jointly yield weak bivalence. But they do not yield the strong bivalence principle (SB), because they do not yield (ME). In fact, one can easily show that if any sentence is a counter-example to (U), then (TDEF1) counts it as true and (FDEF1) counts it as false. Thus if (U) fails for vague sentences in borderline cases, (TDEF1) and (FDEF1) make those sentences both true and false. Such a consequence would generally be regarded as unacceptable.

One or two theorists of vagueness do suggest that a vague sentence is true if and only if at least one of its admissible sharpenings is true, and is false if and only if at least one of its admissible sharpenings is false (Hyde 1997). Such a view is know as *subvaluationism*, by analogy with supervaluationism (discussed below). Subvaluationism resembles the envisaged combination of (TDEF1) and (FDEF1) with the negation of (U). The various things said by a sentence correspond to its various sharpenings. However, subvaluationism has highly counter-intuitive consequences. Consider, for example, 'small' as a vague predicate of natural numbers in a context in which one admissible sharpening puts its cut-off point between n-1 and n while another equally admissible sharpening puts the cut-off point between n and n+1. Then the conjunctions 'n-1 is small and n is not small' and 'n is small and n+1 is not small' both count as true according to subvaluationism, because each is true on at least one sharpening. Moreover, since we might know that both sharpenings were admissible, we might know that both conjunctions were true. Presumably, therefore, we should assert each of them if the question arises. But we are not entitled to assert (rather than stipulate) of any place in particular that the cut-off point for a vague predicate comes there. Still less are we entitled to assert both conjunctions, for they are mutually inconsistent; not even subvaluationism makes their conjunction 'n-1 is small and n is not small and n is small and n+1 is not small' true. Consequently, one should reject subvaluationism. It faces all the main problems of supervaluationism and more besides. In what follows I assume the principle (ME) that truth and falsity are mutually exclusive.

Thus, in the absence of (U), I reject the combination of (TDEF1) and (FDEF1).

One could preserve both (ME) and either one of (TDEF1) and (FDEF1) while rejecting (U) by treating truth and falsity as contradictories. For example, suppose that one accepts (TDEF1). Then to treat falsity as non-truth is to accept the dual of (FDEF1):

(FDEF2) $\forall s \forall c[\mathbf{False}(s,c) \equiv \forall P[\mathbf{Say}(s,c,P) \supset \sim P]]$

A sentence is false if nothing that it says to be so is so. (TDEF1) and (FDEF2) jointly entail the full bivalence principle (SB) itself, and therefore both (ME) and (WB). But since they merely define **True** and **False**, they do not entail (U). Thus they do not jointly entail (T) or (F). In particular, (FDEF2) does not entail (F): although it entails (F→), it does not entail (F←). In any context in which (U) fails, the relevant sentence counts as true by (TDEF1) and as not false by (FDEF2). Therefore, if (U) fails for all vague sentences in borderline cases, all such sentences count as true and not false —and one can know that. Thus the combination of (TDEF1), (FDEF2) and the negation of (U) represents a route to strong bivalence incompatible with that taken by the epistemicist.

A similar route to strong bivalence equally inconsistent with epistemicism combines (FDEF1) with the negation of (U) and a definition of truth as non-falsity. It yields the dual of (TDEF1):

(TDEF2) $\forall s \forall c[\mathbf{True}(s,c) \equiv \forall P[\mathbf{Say}(s,c,P) \supset P]]$

A sentence is true if everything that it says to be so is so. (TDEF2) and (FDEF1) jointly entail (SB), and therefore both (ME) and (WB). Since they merely define **True** and **False**, they do not entail (U) or (T) or (F). In particular, (TDEF2) does not entail (T): although it entails (T→), it does not entail (T←). In any context in which (U) fails, the relevant sentence counts as not true by (TDEF2) and as false by (FDEF2). Therefore, if (U) fails for vague sentences in borderline cases, such sentences count as false and not true —and one can know that.

Although the two combinations just considered secure strong bivalence, they do so at a high cost to classical intuitions. They undermine the natural conception of truth and falsity as somehow parallel notions: one of the defining conditions is quantified universally, the other existentially. Moreover, the presence of just one of (TDEF1) and (FDEF1) suffices to yield some of the counter-intuitive consequences of subvaluationism. For example, given the persuasive principle (N) linking what a sentence says to what its negation says, one can show that in any context in which a sentence is a counter-example to (U), so is its negation: from **Say(s,c,P) & Say(s,c,Q)** one can derive **Say(Ns,c,~P) & Say(Ns,c,~Q)**. Now suppose that (U) fails in a borderline case *n* for the vague predicate 'small'. Then (TDEF1) counts both '*n* is small' and '*n* is not small' as true. That (FDEF2) does not count '*n* is small' as false too is not much consolation, for that is achieved only by not counting a sentence with a true negation as false; (FN) is violated from right to left. Similarly, (FDEF1) counts both '*n* is small' and '*n* is not small' as false. That (TDEF2) does not count '*n* is not small' as true too is equally little consolation, for that is achieved only by not counting the negation of a false sentence as true; (FN) is violated from left to right. Both (TDEF1) and (FDEF1) are to be rejected. What remains is the combination of (TDEF2) and (FDEF2).

Given (TDEF2) and (FDEF2), failures of (U) produce truth-value gaps rather than truth-value gluts. If vague sentences in borderline cases are counter-examples to (U), then such sentences are neither true nor false. This view has an obvious analogy with supervaluationism, on which a sentence is true if and only if all its admissible sharpenings are true, and is false if and only if all its admissible sharpenings are false. As with subvaluationism, the various things said by a sentence would correspond to its various sharpenings.

An immediate problem is that (TDEF2) and (FDEF2) together do not make truth and falsity mutually exclusive, even in the presence of (U). If a sentence says nothing, they count it as vacuously both true and false. Indeed, they count mountains and lakes as both true and false. To overcome this problem, one can add clauses requiring that a sentence is true or false only if it says something:

(TDEF2*) $\forall s \forall c[\mathbf{True(s,c)} \equiv [\exists \mathbf{P\ Say(s,c,P)}\ \&\ \forall \mathbf{P[Say(s,c,P)} \supset \mathbf{P}]]]$

(FDEF2*) $\forall s \forall c[\mathbf{False(s,c)} \equiv [\exists \mathbf{P\ Say(s,c,P)}\ \&\ \forall \mathbf{P[Say(s,c,P)} \supset \mathbf{\sim P}]]]$

This modification preserves the analogy with supervaluationism; the latter does not permit sentences to be vacuously both true and false, for admissible sharpenings are defined at the level of a language as a whole and supervaluationists assume that there is always at least one such sharpening. (TDEF2*) and (FDEF2*) jointly entail (ME), (T→) and (F→); they do not entail (U), (T), (T←), (F), (F←), (WB) or (SB). Since counter-examples to (U) must satisfy the extra clause **∃P Say(s,c,P)**, they still count as neither true nor false by (TDEF2*) and (FDEF2*). If borderline cases falsify (U), they are classified as involving truth-value gaps.

6. THE CASE FOR THE UNIFORMITY PRINCIPLE

In the preceding discussion I assumed, for the sake of argument, that (U) can fail: a sentence may in the same context say something that is so and something that is not so. If so, the supposed failure appears to give succour to supervaluationism. It is time to evaluate the assumption that (U) can fail.

There is a non-technical notion of saying on which to say something can also be to say some of its immediate logical consequences. In that sense, saying that it is cold and wet might involve both saying that it is cold and saying that it is wet; saying that Nenad is absent might involve saying that someone is absent. That notion of saying is clearly irrelevant to the present problem, for it would yield counter-examples to (U) even in unproblematic non-borderline cases. For example, perhaps it is cold but not wet; perhaps Nenad is present but someone else is absent. Since the relevant instances of (T) and (F) should hold in such unproblematic cases, so should their consequences, the relevant instances of (U). Thus we should interpret **Say(s,c,P)** in (T), (F) and (U) to mean something like: s says in c *just* that P. But then how can (U) fail? If, in a given context, a sentence says *just* that P and says *just* that Q, how could the proposition that P be anything other than the proposition that Q? Similarly, if we read **Say(s,c,P)** as something like 'The propositional content of s in c is that P', then the uniqueness implied by the definite article leaves no obvious room for (U) to fail.

Someone might intentionally use physically the same words to say each of two things simultaneously. But it is not *ad hoc* to treat the two intentions as creating two simultaneous contexts of utterance for that sentence, so (U) is still not falsified (see Andjelković and Williamson 2001 for detailed discussion).

Suppose that Harry is a borderline case for baldness. It is natural to hold that the sentence 'Harry is bald' says in the present context that Harry is bald. But, if so, why think that it says something else too? *What* else does it say? If the sentence does not say that Harry is bald, the question arises again: what does it say? 'Harry is bald' seems to say just that Harry is bald. Even the semantic paradoxes do not seem to be counter-examples to the qualified disquotational principle that *if* the sentence «P» says anything at all in this context, it says just that P. Thus it is hard to understand how (U) could fail in the present context. But if it cannot fail in the present context, then it cannot fail in any ordinary context, for there is nothing special about the present context. Since borderline cases such as Harry occur in the present context, vagueness would not falsify (U).

Someone might use the supervaluationist or subvaluationist notion of an admissible sharpening in an attempt to explain a sense of **Say** on which (U) could fail: a sentence would say its admissible sharpenings. Such an explanation is only as clear as the notion of an admissible sharpening. Supervaluationists have great difficulty in giving an adequate account of that notion in such a way that it does not reduce to something epistemic (Williamson 1994: 164, 1995 and 1997: 216-217, the last of which responds to McGee and McLaughlin 1995).

Fortunately, the uniformity principle (U) can be supported by arguments more rigorous than the foregoing remarks. The strategy is as follows. We start with (TDEF2*), a definition of truth of the sort that someone who rejects (U) might accept, and show that under plausible assumptions about compositionality it still leads to (U). Since (TDEF2*) makes **True** behave in a distinctively supervaluationist fashion only if (U) fails, the appearance of support lent by (TDEF2*) to supervaluationism is illusory. I will also show that under plausible assumptions about compositionality (TDEF1) too leads to (U). Thus the appearance of support lent by (TDEF1) to subvaluationism is equally illusory.

The argument assumes that sentences s and t can be connected by a material biconditional into a sentence Est (≡ is the material biconditional in the metalanguage). Although the argument does not apply directly to sentences in a language without such a connective, that is not a serious limitation. For a language with a material biconditional may contain arbitrarily vague sentences, or sentences with any other features that might be thought to undermine (U). Thus if (U) holds for languages with a material biconditional, the relevant theories that predict the failure of (U) are false, and there is no longer any reason to reject (U).

I will now argue from (TDEF2*) to (U). I first note a plausible principle of compositional semantics. Roughly, since E expresses ≡, Est says the biconditional of what s says and what t says:

(E1) $\forall s \forall t \forall c \forall P \forall Q[[\mathbf{Say}(s,c,P)\ \&\ \mathbf{Say}(t,c,Q)] \supset \mathbf{Say}(Est,c,P \equiv Q)]]$

Next, note that if a sentence says something then its biconditional with itself is true:

(E2) $\forall s \forall c \forall P[\mathbf{Say}(s,c,P) \supset \mathbf{True}(Ess,c)]$

Supervaluationists would certainly accept (E2), since every sharpening of Ess is true; it is a classical tautology. A special case of (E1) is:

(1) $\forall s \forall c \forall P \forall Q[[\mathbf{Say}(s,c,P)\ \&\ \mathbf{Say}(s,c,Q)] \supset \mathbf{Say}(Ess,c,P \equiv Q)]]$

Next, recall that (TDEF2*) yields (T→), the half of (T) acceptable to supervaluationists. A special case of (T→) is:

(2) $\mathbf{\forall s\forall c\forall P\forall Q[Say(Ess,c,P\equiv Q) \supset [True(Ess,c) \supset [P\equiv Q]]]}$

From (1) and (2) one has:

(3) $\mathbf{\forall s\forall c\forall P\forall Q[[Say(s,c,P)\ \&\ Say(s,c,Q)] \supset [True(Ess,c) \supset [P\equiv Q]]]}$

But (E2) allows one to discharge the condition **True(Ess,c)** from (3) and thereby derive (U). Once one has (U), one can easily recover (T) from (TDEF2*) and (F) from (FDEF2*). The bivalence principles (SB) and (WB) then follow as before from (T) and (F).

Conversely, one can explain (E2) on general grounds by deriving it from (T) and (E1). For an instance of (T) is $\mathbf{Say(Ess,c,P\equiv P) \supset [True(Ess,c) \equiv [P\equiv P]]}$, which yields $\mathbf{Say(Ess,c,P\equiv P) \supset True(Ess,c)}$, while an instance of (E1) is $\mathbf{Say(s,c,P) \supset Say(Ess,c,P\equiv P)}$; together, these give (E2).

Although (TDEF2*) and (FDEF2*) appear to invite a supervaluationist treatment of vagueness, they do not really do so. Such a treatment would involve the denial of (U). The foregoing argument shows that that in turn would require the supervaluationist to deny (E1). But that is too high a price to pay, for it destroys our conception of what biconditionals say.

There is a similar argument for (U) from (TDEF1), the truth definition with an apparently subvaluationist flavour. The argument also assumes that each relevant sentence s has a negation Ns; this assumption is harmless in the present dialectical situation for a reason just like that already given in relation to the assumption that the relevant sentences are closed under the biconditional function E. In place of (E2), this argument needs:

(E2*) $\mathbf{\forall s\forall c \sim True(NEss,c)}$

(E2*) does not need the condition that NEss says something, for if it says nothing it is certainly not true. Subvaluationists should accept (E2*), since no admissible sharpening of NEss is true; it is a classical contradiction. As before, one derives (1) from (E1). One then applies (N) to (1) to reach:

(1*) $\mathbf{\forall s\forall c\forall P\forall Q[[Say(s,c,P)\ \&\ Say(s,c,Q)] \supset Say(NEss,c,\sim[P\equiv Q])]]}$

Next, recall that (TDEF1) yields (T←), the half of (T) acceptable to subvaluationists. A special case of (T←) is:

(2*) $\mathbf{\forall s\forall c\forall P\forall Q[Say(NEss,c,\sim[P\equiv Q]) \supset [\sim[P\equiv Q] \supset True(NEss,c)]]}$

From (1*) and (2*) one has:

$$(3^*)\forall s\forall c\forall P\forall Q[[\mathbf{Say}(s,c,P)\ \&\ \mathbf{Say}(s,c,Q)] \supset [\sim[P\equiv Q] \supset \mathbf{True}(NEss,c)]]$$

But (E2*) allows one to apply *modus tollens* to the final conditional in (3*) and thereby derive (U). Once one has (U), one can easily recover (T) from (TDEF1) and (F) from (FDEF1). The bivalence principles (SB) and (WB) then follow as usual from (T) and (F). Although (TDEF1) and (FDEF1) appear to invite a subvaluationist treatment of vagueness, they do not really do so. Such a treatment would involve the denial of (U). The argument shows that that would require the subvaluationist to deny (E1). As before, that is too high a price to pay.

The crucial premise in both arguments for (U) is (E1). It articulates the natural way in which a biconditional sentence says something when its constituent sentences say something. One might come to doubt (E1) by equating saying something with not clearly not saying it. This can be formalized in terms of the 'clearly' operator Δ. Given that the antecedent of (E1) is clearly sufficient for its consequent, one has:

$$(E1\Delta)\ \forall s\forall t\forall c\forall P\forall Q\ \Delta[[\mathbf{Say}(s,c,P)\ \&\ \mathbf{Say}(t,c,Q)] \supset \mathbf{Say}(Est,c,P\equiv Q)]]$$

But from (E1Δ) one cannot infer:

$$(E1\Delta\Delta)\ \forall s\forall t\forall c\forall P\forall Q[[\sim\Delta\sim\mathbf{Say}(s,c,P)\ \&\ \sim\Delta\sim\mathbf{Say}(t,c,Q)] \supset \sim\Delta\sim\mathbf{Say}(Est,c,P\equiv Q)]]$$

Similarly, when $\Box$ and $\Diamond$ mean necessity and possibility respectively, $\Box[[\mathbf{P}\ \&\ \mathbf{Q}] \supset \mathbf{R}]$ does not generally entail $[\Diamond\mathbf{P}\ \&\ \Diamond\mathbf{Q}] \supset \Diamond\mathbf{R}$. For example, if **P** expresses a contingency then $\Box[[\mathbf{P}\ \&\ \sim\mathbf{P}] \supset [\mathbf{P}\ \&\ \sim\mathbf{P}]]$ is trivially true while $[\Diamond\mathbf{P}\ \&\ \Diamond\sim\mathbf{P}] \supset \Diamond[\mathbf{P}\ \&\ \sim\mathbf{P}]$ is false. For analogous reasons, if it is not clear that s does not say that P and not clear that s does not say that Q, it does not follow that it is not clear that Ess does not say that P if and only if Q. It may be clear that Ess does not say that P if and only if Q just because it is clear that P and not Q. But if it is clear that Ess does not say that P if and only if Q, then that is so because it is clear that s does not *both* say that P *and* say that Q. Thus (E1) and (E1Δ) themselves are unthreatened. Saying something should not be confused with the more complex matter of not clearly not saying it. There is a natural and theoretically central notion of saying which satisfies (E1), (E1Δ) and (U). That is the notion needed in an account of truth and falsity. If one so wishes, one can then introduce a secondary notion of quasi-saying by the formula **~Δ~Say(s,c,P)**, but quasi-saying something does not amount to saying it. One might for a moment be tempted to suppose that quasi-saying is a more precise notion than saying, because it takes vagueness explicitly into account with its use of the clarity operator. However, the phenomenon of higher-order vagueness implies that quasi-saying is a vague notion too, just like saying. Unlike saying, quasi-saying has no special theoretical significance.

We have examined attempts to reject the uniformity principle (U) and therefore the principles (T) and (F) about truth and falsity in favour of definitions of truth and falsity which, in the spirit of supervaluationism or subvaluationism, yield only one or other half of (T) and (F). Those attempts violate independently plausible constraints of semantic compositionality. Once the uniformity principle is restored to its rightful place, those definitions yield (T) and (F) after all. Thus, given classical logic, they also yield the principle (SB) of strong bivalence. The apparent loophole in the classical case for bivalence is therefore merely apparent.

ACKNOWLEDGEMENT

This paper draws extensively on my contribution to Andjelković and Williamson 2001. I am very grateful to Miroslava Andjelković for her part in that joint project. That paper contains amplifications of some of the arguments in this one.

REFERENCES

Andjelković, Miroslava. (1999). 'Williamson on bivalence'. *Acta Analytica* 14 (issue 23): 27-33.

Andjelković, Miroslava, and Williamson, Timothy (2000). 'Truth, falsity and borderline cases'. *Philosophical Topics* 28: 211-244.

Fine, Kit. (1975). 'Vagueness, truth and logic'. *Synthese* 30: 265-300. Reprinted in Keefe and Smith (1996).

Hyde, Dominic. (1997). 'From heaps and gaps to heaps of gluts'. *Mind* 106: 440-460.

Keefe, Rosanna, and Smith, Peter, eds. (1996). *Vagueness: A Reader*. Cambridge, Mass. and London: MIT Press.

McGee, Vann, and McLaughlin, Brian. (1995). 'Distinctions without a difference'. *Southern Journal of Philosophy* 33 (supplement): 203-251.

Williamson, Timothy. (1994). *Vagueness*. London and New York: Routledge.

Williamson, Timothy. (1995). 'Definiteness and knowability'. *Southern Journal of Philosophy* 33 (supplement): 171-191.

Williamson, Timothy. (1997). 'Imagination, stipulation and vagueness'. In E. Villanueva, ed., *Philosophical Issues 8: Truth*. Atascadero CA: Ridgeview.

Williamson, Timothy. (1998). 'Indefinite extensibility'. *Grazer Philosophische Studien* 55: 1-24.

Williamson, Timothy. (1999). 'Andjelković on bivalence: a reply'. *Acta Analytica* 14 (issue 23): 35-8.

Chapter 4

MEANING FINITISM AND TRUTH[1]

Martin Kusch
Cambridge University

1.

'Meaning finitism' is a theory of linguistic meaning that has been developed since the late seventies by Barry Barnes and David Bloor.[2] Barnes and Bloor are the two founding fathers of the 'Edinburgh School' in the Sociology of Scientific Knowledge. In this paper I seek to reconstruct and defend meaning finitism by exploring its consequences for our understanding of truth.[3] Given the constraints of time, I will here only be able to scratch the surface of the

[1] A much earlier version of this paper was discussed at the Moral Sciences Club in Cambridge in 1998. I am grateful to Anjan Chakravartty, Michael Esfeld, Anandi Hattiangadi, Jane Heal, Susan James, Tim Lewens, Peter Lipton and Hugh Mellor for many objections and suggestions on that occasion. I have also profited from comments by David Bloor, David Chart, Jeremy Gray, Matthias Hild, Matthew Ratcliffe and Paul Teller.

[2] The best general accounts are Barry Barnes, David Bloor, John Henry, *Scientific Knowledge: A Sociological Analysis* (London: Athlone Press, 1996), Chapter Three; and D. Bloor, *Wittgenstein, Rules and Institutions* (London: Routledge, 1997). See also B. Barnes, 'On the Extensions of Concepts and the Growth of Knowledge,' *Sociological Review* 30 (1982), 23-44.

[3] I have used and developed Barnes' and Bloor's views on meaning in my earlier work. See M. Kusch, *Psychological Knowledge: A Social History and Philosophy* (London: Routledge, 1999); Harry Collins and M. Kusch, *The Shape of Actions: What Humans and Machines Can Do* (Boston: M.I.T. Press, 1998).

K. Korta and J. M. Larrazabal (eds.), Truth, Rationality, Cognition, and Music, 67-83.

issue. I hope to convince you that I am at least scratching in the right places, and in the right direction. (I dig deeper in a recently published book.[4])

In order to get an initial foothold in meaning finitism—and its opposite, 'meaning determinism'—it helps to recall a theological distinction between two ways in which God might be, or might have been, creating the world. According to 'deism,' God initially creates the world but does not subsequently intervene in it. After its creation, the world runs its predetermined course. According to the 'continued-creation' view, however, creation and preservation of the created order are one and the same process. In other words, the work of creation is never done; creation is a process that continues from moment to moment.[5]

Replace 'world' with 'meaning' and 'God' with 'humans' and you have the needed analogy. The deistic view of creation has its analogue in meaning determinism. For simplicity's sake, let me focus on general terms only. Many philosophers of language work with the following idealisation of linguistic practice. They assume that speakers have—always already—settled on the meanings of their terms. Philosophical reflection only focuses on problems that temporally follow this initial determination. Put differently, the creation or constitution of meaning is treated as a *resource* rather than as a *topic* of philosophical reflection. The topic of philosophical-semantic reflection lies elsewhere: its aim is to understand how the previously constituted meaning *determines* an extension, and how the term is *true of* this extension. It is this determination and this relationship of truth that 'deistic philosophy of language,' or meaning determinism, seeks to conceptualise.

The continued-creation view in the philosophy of language—that is, meaning finitism—refuses the mentioned idealisation. Meaning finitism rejects the idea that meanings determine use; instead it sees meanings as the continuously created product of verbal behaviour. Meanings are the continuously created outcomes of the use of words, not the determinants of that use. Since meanings are continuously made (and re-made) by language users, these meanings are never sufficiently stable and fixed for them to be able to determine extensions. Accordingly, it does not make sense either to claim that terms are true of their extensions. Meaning finitism is led to downgrade the importance of truth in semantics.

[4] M. Kusch, *Knowledge by Agreement: The Programme of Communitarian Epistemology* (Oxford: Oxford University Press, 2002).

[5] I am grateful to Stephen Snobelen for helpful email exchanges regarding this distinction.

2.

I now turn to a more detailed exposition of meaning finitism. To start us off, let me first formulate an individualistic version. According to this version, someone, say John, can be said to know the meaning of a linguistic expression, say 'chair,' if and only if he meets the following conditions. First, John has been trained to apply the word 'chair' to entities of a certain kind; and such training has taken place in the presence of a finite number of examples, or, as I shall say slightly more technically, a finite 'array of exemplars.' Second, on the basis of this training, John has developed a disposition to apply 'chair' to some entities he encounters but not to others. Third, the application of 'chair' to a newly encountered entity involves, or can be reconstructed as, a judgement of similarity concerning the newly encountered entity and one or more of his exemplars. Fourth, an entity that John judges to be similar to one or more of his exemplars can itself become a further exemplar. Fifth, any element of the original array of exemplars may later be dropped. And sixth, similarity is in the eye of the beholder. Which entities John regards as similar is determined by his earlier judgements of similarity, as well as by his interests, goals, and commitments.

Note that finitism runs counter to the idea that terms have fixed extensions. If finitism is correct then it is not possible to say that the extension of 'chair' consists of everything that might have been, or might be, truly classed as a chair. The content of the class 'chair' depends on a stream of decisions over time.

3.

Like Barnes and Bloor, I combine meaning finitism with meaning collectivism; that is, I hold that meaning is a social-collective phenomenon. In its Edinburgh version, meaning collectivism has a more familiar aspect and a somewhat less well-known side. The more familiar aspect is Wittgenstein's argument against the possibility of a private language.[6] Put in a nutshell, meaning is, for the individual, a normative phenomenon; normativity presupposes an independent standard; and this independent standard is set by the consensus of the individual's community. It is negotiation within, and the consensus of, the relevant linguistic community that determines which similarities between exemplars and newly

[6] Ludwig Wittgenstein, *Philosophical Investigations*, transl. G. E. M. Anscombe (Oxford: Blackwell, 1967).

encountered entities are salient and constitute correct applications of the expression. I shall take this Wittgensteinian argument for granted here.

The less well-known side of meaning collectivism is Bloor's claim that meanings are social institutions.[7] According to the 'performative theory' of social institutions—first suggested by Barnes, and more recently developed further by Bloor and myself[8]—a social institution is a self-referring system of talk, beliefs and actions. Take the social institution of money. What makes a metal disc a coin, that is, an instance of money? Nothing more nor less than that a collective treats the metal disc as money, that it talks of it as money, that it uses it as money, and that it sanctions and enforces this talk and this use. Money, we might say, is what we collectively take to be money; money is a social *status* that we collectively impose upon certain objects. Put differently, money has no form of existence outside the practices of talking about money, reminding people of the use of money, teaching children what is money, exchanging money for various goods or other kinds of money, etc.

Note the following two important features of social institutions: First, institution-creating and -recreating talk is ultimately self-referring. It is the talk about money that creates money as the referent for such talk. And second, social institutions are subject to the 'logic' of finitism. They are performed, and their paths are not predetermined by rules and norms. Social actors have to make decisions regarding metal disks and other objects: Is this a situation in which it is appropriate to use metal disks for exchange? Is this metal disk similar to those we have used as money before?

We can now see why meanings are social institutions. First, since linguistic meaning is—as we have learnt from Wittgenstein—a social phenomena, it cannot be reduced to what exists in the heads of individuals. Individuals cannot constitute normativity unless they interact. Second, arrays of exemplars—*qua* precedents for the application of words—are social institutions. Being an exemplar is a social status, and the social status of 'being an exemplar' (or whatever the more colloquial expressions for this

[7] Bloor, Wittgenstein, *Rules and Institutions*, Ch. 3.

[8] B. Barnes, "Social Life as Bootstrapped Induction," *Sociology* 17 (1983): 524-45; idem, *The Nature of Power* (Cambridge: Polity Press, !988); idem, *The Elements of Social Theory* (London: UCL Press, 1995); D. Bloor, "Idealism and the Sociology of Knowledge," (unpublished manuscript, 1996); idem, *Wittgenstein, Rules and Institutions*; Kusch, *Psychological Knowledge*. Compare also John R. Searle, *The Construction of Social Reality* (London: Allen Lane, 1995); Esa Itkonen, *Grammatical Theory and Metascience: A Critical Investigation into the Methodological and Philosophical Foundations of 'Autonomous' Linguistics* (Amsterdam: Benjamins, 1978); Eerik Lagerspetz, *The Opposite Mirrors: An Essay on the Conventionalist Theory of Institutions* (Dordrecht: Kluwer, 1995); Raimo Tuomela, *The Importance of Us: A Philosophical Study of Basic Social Notions* (Stanford, Cal.: Stanford UP, 1995).

status happen to be) exists only in exemplar-talk or meaning-talk. Third, the closest we can get to identifying meanings, is to identifying socially sanctioned exemplars and dispositions. Meaning thus has no existence outside of practices in which it is invoked, ascribed, corrected, challenged, and conveyed.

Compared with our initial individualist version of finitism, a collectivist reading of finitism will thus insist on two additional points: one, that respects of similarity are determined by *collective* rather than *individual* interests, goals or commitments; and two, that one knows the meaning of an expression only insofar as one is a participant in social interactions in which a consensus concerning that meaning emerges for a time.

Lest this position sound implausible or excessively vague, I had better something to say on how such consensus emerges and is sustained. It is important to keep in mind that very different models of consensus-formation can be distinguished. Here is my own simple-minded attempt at a classification, based on a mechanical model.[9] Assume we have a large population of clocks, each one with its own characteristic, individual, way of running faster and slower than most other clocks. How can we make these clocks agree, that is, reach consensus on one time? A first way is to have them all connected up to one 'master-clock' that regularly resets all clocks to its time. A second way is to have all clocks linked up to an average-calculating and clock-resetting device. This device calculates the average time on the basis of the times 'send in' by all individual clocks, and resets them to this time. The third and perhaps most interesting model is this. Assume that the clocks move about in a confined space, and that each one of them regularly collides with one or more of the others. Whenever such collisions happen, the colliding clocks 'collectively' calculate the average of their times, reset themselves to that time, and then separate—each one now again displaying its individualism—before they collide with other clocks. In this scenario all consensus is local and temporary, and none of the clocks can 'know' what the 'collective consensus'—that is, the bandwidth of all times—of all clocks would be. This strikes me as a good analogy of social life as far as consensus over linguistic matters, that is, meanings, is concerned.

[9] Kusch, *Psychological Knowledge*, pp. 270-272.

4.

I have said little about 'true' thus far, but it should not be difficult to see that collectivist finitism does not just apply to predicates like 'chair' but also to expressions like 'true' or 'false'. Let me mention a few examples.

Think of how we learn the words 'true' or 'false'. We get shown (or explained) a number of situations where we can predicate 'true' or 'false' of people ('true friends', 'false friends', 'true villains'), materials ('true gold', 'false gold'), and statements ('It is true [or false] that Mary is the head of the physics department'). On the basis of such learning situations we develop dispositions to use the word 'true' in some circumstances but not others. We do so ultimately on the basis of—explicit or tacit—judgements of similarity with earlier situations where we used 'true' correctly. Whether our judgements are acceptable, whether we can, as it were, get away with making them, depends on the judgements of others.

'Is this piece of gold like the piece of true and pure gold that John showed me yesterday? Yes. Well, I guess it's a piece of true gold then.'—'Does John behave towards me like Mary does? Pretty similarly, he too helped me to write this paper. Well, I guess John is a true friend too then.—Does the statement 'Mary is the head of the physics department' fit what I know about this place in a similar fashion as does the statement 'John is head of the philosophy department'? Yes. Well, I guess the statement about Mary is true then.'

In all of these cases we can of course imagine negotiations over the application of 'true'; someone, say Peter, might deny that John's behaviour towards me is sufficiently similar to that of Mary, and Peter might have a social interest in doing so. Or someone might wish to challenge that a statement like 'Mary is the head of the physics department' relates to the world in the way a sentence like 'The cat is on the mat' does. Such 'someone' might therefore deny that a statement like 'Mary is the head of the physics department' is true.

Thus, meaning collectivism and meaning finitism apply to 'true', 'false, 'fact', and 'correspondence' just as meaning collectivism and finitism apply to any other word. They too are learnt on the basis of a finite number of examples, and they too can be used correctly or incorrectly only against the backdrop of a linguistic community. Moreover, and perhaps most importantly, if, or since, 'true' and 'false' behave like 'chair', then they too do not come with a fixed extension. 'True statement' does not have an extension of all statements that might have been, or might be, truly classed as true. Which statements will be classed as true depends on how we collectively decide to class statements; it depends on how we collectively decide to develop similarities between past exemplars and new statements.

5.

Finitism about truth has a number of important philosophical consequences. Relativism concerning truth is one of them.[10] After all, according to my account, the exemplars of truth as well as the relations of similarity, are selected under social, contingent conditions, influenced by social-political concerns. There are no *absolutely* right or wrong ways for picking exemplars of truth and judging the similarity between earlier exemplars and new instances. And thus there are no 'facts' that force us to pick and develop exemplars and respects of similarity one way rather than another. Different possible or real communities may develop their arrays of exemplars—say of true friends or true statements—in many different ways, and there is no vantage point from which communities can be judged right or wrong; no vantage point that is, other than the local and contingent vantage point of some given community.

Moreover, the finitist account of truth implies that notions like 'the whole truth' or 'progress towards *the* truth' are obscure. This should not occasion surprise given that finitism rejects the idea that 'true' has a fixed extension. But it is perhaps worth going over this ground in more detail anyway.

To begin with, in my account 'the whole truth' must be defined with respect to a community; to speak of truth independently of a specific community's exemplars and dispositions makes no sense on the basis of what I have laid out above. Moreover, a community would only have reached 'the whole truth' if it met the following conditions: (i) it has arrived at a stable consensus view concerning an array of truth-exemplars; (ii) it agrees on which statements are similar to the exemplars; (iii) it has reached consensus that no other statement is similar to any of the truth-exemplars; and (iv) whatever further goals and interests will emerge in this society, the sense of similarity and analogy will never ever change again.—No doubt, this is a most unlikely scenario.

In fact, it is incoherent to make a claim like the following (*):

> (*) We can imagine that there is/were a set of propositions (statements, sentences) such that this set is independent of our (linguistic) practices; such that these propositions (statements, sentences) are true or false independently of us; and such that the progress of inquiry means that we can get our hands (minds) on ever greater numbers of the true propositions (statements, sentences) of this set.

[10] Barnes and Bloor have defended epistemological relativism in a number of places. See e.g. 'Relativism, Rationalism and the Sociology of Knowledge,' in *Rationality and Relativism*, ed. Martin Hollis and Stephen Lukes (London: Routledge, 1982), pp. 21-47. I go beyond them here in making explicit the implications of meaning finitism for relativism and truth.

Given finitism about 'true', this claim is incoherent since there cannot ever be such a set prior to our deciding on just what we are to take as the analogies and differences, similarities and dissimilarities, between past and future occasions of applying 'true' and 'false'. Since the future is open, the application of 'true' is open, and since the application of 'true' is open, 'truth' is open. (*) is of course again based on the mistaken assumption according to which 'true statement' has a fixed extension.

But could one not say that we might get ever closer to the truth by the lights of our current understanding of 'true'? Such that if —counterfactually— we were transported with our exemplars and dispositions into a future state of science, we would be able to see that future science is closer to the truth than us now? No. This will not do. Assume we were to judge the future science as closer to the truth than our own. Surely such judgement would still be a judgement shaped and determined by our current sense of similarities and analogies, a sense shaped by our current goals and interests. And thus such judgement would be internal, and relative, to our own practices.

A third consequence of truth finitism concerns the role of truth as a resource in philosophical semantics. For the deistic meaning determinist the key idea regarding truth is that terms are true of their extension. And this idea is closely related to the claim that meanings determine extensions. What closes the triangle is that meanings are associated with terms. The traditional point of view thus involves something like the following picture (Picture 1):

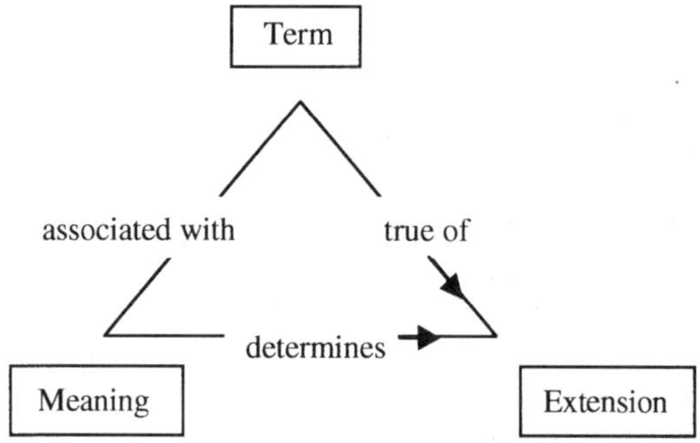

Picture 1. The meaning determinist's triangle

Meaning finitism challenges this picture. It denies the existence of stable meanings that determine extensions, and it rejects the idea that meanings are stable entities that can be associated with terms. And this also affects the relation between term and extension: There is nothing that terms are true of prior to the step-by-step decisions of communities of speakers. Indeed, the same holds of 'true' itself. There are no extensions over and above the ever-changing and developing arrays of exemplars.

6.

Needless to say, meaning finitism and truth finitism run counter to many of our pre-philosophical and philosophically-educated intuitions about meaning and truth. And hence objections abound. I shall spend the reminder of this paper trying to answer some of the most obvious ones, proceeding from worries about meaning finitism in general to worries about truth finitism in particular.

First Objection: Meaning finitism is wrong to assume that the application of terms involves judgements. When (on a sunny day) I call the sky 'blue,' my application of the concept 'blue' is not based on any similarity judgement. To say 'the sky is blue' is a judgement of course; but the meaning finitist is committed to saying that this judgement is based on a series of others: judgements concerning the similarity between this sky and earlier skies, this blue and earlier blues and so on. – *Reply*: It is true that most of the time we are content to let our unreflective 'just doing it' constitute proper usage. And yet—as pointed out by Barnes,

> ... it is just being content in such circumstances which makes concept application in those cases plausibly describable as a matter of judgement. In those cases we allow, *as we need not*, our unreflective linguistic activity to develop proper usage: we place our confidence, *as we need not*, in our habits and routines. It is true that if we did not trust ourselves, and our own routine linguistic competences, most of the time, the use of language would become impossibly difficult and discourse would cease. But such trust in what comes routinely and unreflectively is essential in no *particular* case, so that its presence in any particular case can be said to reflect a contingent judgement that usage may be best developed in that way in that case.[11]

Second Objection: Meaning finitism is no news. It is just a restatement of meaning scepticism as formulated by Saul Kripke in his book on Wittgenstein.[12] – *Reply:* Undoubtedly, there are important connections

[11] Barnes, 'On the Extensions of Concepts,' p. 34.
[12] Saul Kripke, *Wittgenstein on Rules and Private Language* (Oxford: Blackwell, 1982).

between meaning scepticism and meaning finitism. Bloor has written extensively on the parallels and differences;[13] (and I am myself in the early stages of writing a book on the Kripke debates). Suffice it here to make two observations. First, meaning finitism is not a form of communal dispositionalism. Whether I mean addition by 'plus' is not determined by community-wide homogeneous, or interacting heterogeneous, semantic dispositions. It is negotiated (and possibly renegotiated at a future date) between individuals of my community; and this negotiation is determined by the interacting individuals' semantic dispositions, their goals and interests. Faced with the sceptical challenge regarding 'plus,' the meaning finitist answers as follows: By '+' I mean addition rather than quaddition or skaddition because members of my linguistic community routinely call people who calculate like I have done in the past 'adders;' members of my linguistic community routinely allow people who calculate like I do to call themselves 'adders;' and members of my linguistic community regard as strange people who present themselves as quadders and skadders. Second, the meaning finitist agrees with the meaning sceptic that semantic facts do not determine meaning. But this is not because, as the meaning sceptic would have it, there are no semantic facts. It is rather because semantic facts are, as it were, forever in the making. Facts about meaning are constantly constituted and reconstituted. And precisely because they are constantly (re-) made in and through interaction between individuals, these facts cannot act as determinants of action.

Third Objection: Meaning finitism boils down to saying that all concepts are vague. – *Reply:* It is tempting but incorrect to summarise the finitist theory of meaning as 'all concepts are vague.' The distinction between vague and non-vague stands orthogonal to the distinction between meaning as 'finite' and meaning as 'fixed by extensions.' Take a concept like 'bald,' the prototype of a vague predicate. What makes 'bald' vague is not that its application is based on exemplars and judgements of similarity. Instead, 'bald' is vague because we are collectively willing to accept both '*X* is bald' and '*X* is not bald' as assertible of the same *X* and at the same time. Regarding non-vague terms (like 'one-meter-long') no such tolerance is generally displayed. More formally, a concept *C* is vague if the communal practices of evaluation allow for different speakers to apply both *C* and *non-C* to a given entity. A concept *C* is non-vague if the communal practices of evaluation do not allow for different speakers to apply both *C* and *non-C* to a given entity. Put differently, all acts of concept-application are underdetermined by past practice. But this does not make the concepts

[13] Bloor , Wittgenstein, Rules and Institutions, esp. Ch. 5.

vague. 'Vague' and 'precise' are two social statuses that communities attach to concepts.[14]

Fourth Objection: Meaning finitism conflates words and things. The finitist cannot explain why, in order to correctly classify some animals as 'cats' we should engage in biological enquiry rather than in a semantic study of our linguistic community. Meaning finitism is saying something like "the community can make anything a cat by calling it 'cat'." Hence the only way to find out whether anything is a cat is to see whether the community is calling it so. – *Reply:* This is too quick. The sentence 'the community can make anything a cat by calling it "cat"' is itself based on a conflation: the 'make' is ambiguous between creation and categorization. Communities cannot (physically) create cats by labelling certain furry animals 'cats.' But they can of course group together certain animals by using the label 'cat.' For me (as a member of an English-speaking community), to find out whether a given furry animal is a cat is to primarily attend to the animal; to feel its fur; to listen to the sounds it produces; and to see whether it chases mice. In all this I can keep questions of (communally established) exemplars in the background. This is because I know it to be unlikely that I am out of harmony with my fellows' categorizations of cats. And yet, every act of categorization is a stab into the semantic dark; it involves an evaluation of similarity between exemplars and the new case. And it might always turn out that my fellow speakers' do evaluate this similarity differently from me. Thus there is no guarantee that my act of categorization is right until the others have agreed with it. Put differently, it is true that the meaning finitist cannot make a sharp distinction between questions of empirical truth and questions of semantic correctness. But to deny that these two types of questions can be sharply separated is not of course to be guilty of collapsing the former into the latter.

Fifth Objection: Finitistic truth relativism conflates knowing the truth with knowing the meaning of the word 'true'. Surely, knowing the latter does not mean knowing the former.—*Reply:* The initial force of the objection derives from the tacit assumption that 'true' has a fixed extension; an extension that exists independently of how we collectively decide to develop analogies between past and present uses. Only if the extension is independent of these decisions does it make sense to oppose knowing the truth to knowing the meaning of true. But this is precisely what finitism does not allow us to do. Of course, in everyday life we do distinguish between knowing the meaning of 'true' and knowing truths. My account does not rule against this. In everyday life we assume that someone knows what the word

[14] My comments on vagueness are indebted to Michael P. Lynch, *Truth in Context: An Essay on Pluralism and Objectivity* (Cambridge, Mass.: MIT Press, 1998), p. 61.

'true' means if they can competently participate in negotiations over the 'truth-status' of friends, materials, or statements. One can be a competent participants in such negotiations without knowing more than just the standard truth-exemplars. And obviously one does not need to know what all will, or could, be judged to be similar to these exemplars.

Sixth Objection: Finitistic truth relativism disregards the distinction between 'true' and 'taken for true'. Our picking of exemplars and our deciding on analogies and similarities is all to do with 'taking for true;' but it has nothing to do with 'true' or 'truth' itself. – *Reply:* The distinction between 'merely believed to be true' and 'true' is an important distinction *internal to* our very practice of truth-talk. Presumably, we would not, in general, think that a society had anything like a concept of truth unless it drew a contrast between 'merely believed to be true' and 'true'. But the contrast itself is again subject to the 'logic' of finitism. The contrast between 'true' and 'believed to be true' is one that is learnt on the basis of exemplars; it is thus a contrast that is defined by a consensus-based, changing set of cases. This is what anyone who finds the sixth objection tempting will tend to overlook: He will tacitly assume that while 'believed to be true' is subject to finitism, 'true' has a fixed extension. But that is wrong.

Seventh Objection: Truth finitism is just a rehash of pragmatist theories of truth. – *Reply:* Let me consider James' pragmatism first.[15] James held that truth must be defined in terms of the useful, and spoke of us—us *qua* individuals—as being able to 'create' truths. At the same time, James occasionally endorsed the correspondence theory of truth. James' commentators and critics have rejected the equation of the useful with the true, and have deplored the apparent inconsistency between the utility theory and the correspondence they.[16]

It seems to me, however, that James' position can be partially vindicated, and made sense of, in the light of the collectivistic-finitistic theory presented above. James was right to insist on the importance of the useful. This is because considerations of utility—as fixed by social interests and goals—enter into the negotiations over what are to count as relevant similarities and analogies between truth-exemplars and suggested truth-candidates. No consideration of utility, no application of 'true' to new cases.

James was also right to say that we 'make' or 'create' truths. After all, we decide which similarities are to count as truth-making, and which do not. This choice is not forced upon us by the nature of the things talked about.

[15] William James, *Pragmatism and The Meaning of Truth*, intr. by A. J. Ayer (Cambridge, Mass. and London, England: Harvard UP, 1975).

[16] See, e.g. Bertrand Russell, "William James's Conception of Truth," in *Truth*, ed. by Simon Blackburn and Keith Simmons (Oxford: Oxford UP, 1999), pp. 69-82.

Nor was James incoherent when stressing the importance of correspondence alongside utility. We have our exemplars for what it is for statements to 'fit with,' or 'correspond to' other statements, or the world. And some of these exemplars are shared between the array of exemplars on the basis of which we learn 'true', and the array on the bases of which we learn 'fit'. Interests are involved in judgements of similarity based upon these exemplars. This is not to deny, however, that these judgements are concerned with correspondence or fit.

Nevertheless, James did go wrong in one or two important respects. On the one hand, he erred in thinking that 'true' could somehow be 'defined' in terms of the useful. That considerations of utility figure in the determination of how to apply 'true' to new cases does not mean that 'true' means 'useful'. On the other hand, and as first pointed out by Durkheim[17], James remained an individualist at heart. The creation of truth was at bottom a creation by individuals. James overlooked the objective nature of truth, or as we might say more precisely, James overlooked that accepted truths have the status of social institutions. Truths are things one ought to accept.

Other pragmatist theories of truth, say the consensus theories of Peirce or Habermas[18], were, or are, more collectivistic than James. They err, however, by turning a condition of the possibility of correctness in the application of 'true' into the definition of 'true'. There cannot be a correct application of the word 'true' unless there is a society that collectively constitutes exemplars. But these exemplars do not guide us in the application of the expression 'what is agreed amongst us', or 'what we all ideally should agree with;' they guide us in the application of the expression 'true'. Of course, Peirce and Habermas also go wrong in insisting on an ideal consensus, of science in the limit, or of all rational people at all times. The members of Habermas' ideal speech community are purified to such extent that they lack social interests; and thus they have no way of fixing respects of similarity.

Eighth Objection: Truth is disquotation.[19] And to recognise this is to see that finitism cannot be correct. If truth is captured by the disquotational scheme—

(DS) 'X' is true if and only if p

[17] Emil Durkheim, *Pragmatism and Sociology,* transl. by J.C. Whitehouse, ed. & intr. by J. B. Allcock, with a preface by A. Cuvillier (Cambridge, England: Cambridge UP, 1983).

[18] For a very helpful account of Peirce on truth, see Richard L. Kirkham, *Theories of Truth: A Critical Introduction* (Cambridge, Mass.: MIT Press, 1992), pp. 79-87. For Habermas on truth, see his 'Wahrheitstheorien,' in Jürgen Habermas, *Vorstudien und Ergänzungen zur Theorie des kommunikativen Handeln* (Frankfurt am Main: Suhrkamp, 1984), pp. 127-83.

[19] W.V. Quine, "Philosophy of Logic," in *Truth*, ed. Blackburn and Simmons, pp. 144-46.

—then 'true' does have a definite and determinate content after all. – *Reply:* Meaning finitism resurfaces immediately since DS does not determine to which sentences it ought to be applied. Note only that DS does not apply straightforwardly in cases of vagueness and to Liar-paradox sentences.[20] If 'tall' is a vague predicate, then the truth-value of 'Jim is tall' is indeterminate. And hence 'Jim is tall' does not imply '"Jim is tall" is true.' Or assume that we are dealing with a sentence L with the content: L is false. Applying DS to L we get:

(a) "L is false" is true if and only if L is false.

Since L is the sentence "L is false" we can replace the last "L" in (a) with "L is false." And since double negation equals affirmation we get the unacceptable result:

(b) "L is false" is true if and only if L is true.

The moral should be obvious enough. We must decide case by case whether or not a given sentence is such that we regard the disquotational scheme as applicable. And this involves us in collective decisions over which predicates we wish to regard as vague, and which sentences we regard as paradox-prone.

Ninth Objection: But surely truth exemplars have not really changed over time. Are not Aristotle's exemplars for 'true' pretty much the same as, or highly similar to, our own truth exemplars today?—*Reply:* The answer is 'yes' if we remain in the realm of everyday life and common-sense object. Indeed our coinciding with most other cultures, past or present, in this area is what allows us to speak of them possessing an expression for 'true' in the first place. But of course, the moment we leave that common area to talk about science, religion, or metaphysics differences abound.

7.

In the above, I have tried to explain a theory of meaning and truth that has not, to date, received much attention from cognitive scientists. I believe, however, that it is a theory that cognitive scientists should find congenial. An empirical field like cognitive science cannot afford to make idealisations

[20] See e.g. Mark Sainsbury, "Philosophical Logic," in *Philosophy 1: A Guide through the Subject*, ed. A. Grayling (Oxford: Oxford University Press, 1998), pp. 61-122, here pp. 105-108.

of the form favoured by meaning determinism; it cannot afford to think of the creation of meaning on the deistic model. An empirically adequate theory of meaning must give proper heed to the fact that meaning is continuously created and re-created in real time.

APPENDIX

Some interlocutors have found the following 'game' versions of meaning finitism and meaning determinism helpful in understanding what is at stake. We can model the two different views of meaning on two simple game-scenarios. I call these games 'RISTO' and 'SEPPO.'[21]

Imagine the following game, called 'RISTO.' (Don't try this at home!) In order to play the game one needs a room filled with various objects, two players (A and B) and a stamp (along with an ink pad). During the first stage of the game, player A stays outside the room. While A is out, B takes the stamp, walks around the room and stamps various objects. Some of the stamp-patterns will be openly visible, others will be on object-surfaces that are blocked from sight or covered. All of the stamp-patterns are identical in shape and color. After some time, A is allowed back inside the room. His task is to identify all those objects that have the stamp-pattern; such objects are called 'ristos.'[22]

Our second game, 'SEPPO,' needs the same kind of room as does RISTO. But SEPPO requires three players (A, B, and C). No stamp or ink pad are needed. While A again is out of the room, B and C agree amongst themselves to call three objects 'seppo.' They must choose three objects that they regard as similar. Let us call these objects 'object 1,' 'object 2,' and 'object 3' and their collection the 'seppo-array.' Then A is allowed to return to the room. A is told which three objects are called 'seppo.' Subsequently, A must go around the room, suggest further 'seppos,' and justify his choice to B and C. That is to say, he must find further objects that in his opinion are arguably similar to one or more of the three objects originally picked by B and C. Whether or not A's choice is correct or not depends on the judgments of all three players. When A's suggestion concerning a new seppo is accepted by majority rule (A, B and C all vote), then the new object is entered as object 4 into the seppo-array. At the same time, object 1 is dropped from the array. The game then continues relative to the new, changed seppo-array. After a given number of rounds, B (and later C) continues in A's role, but no new array is chosen at that time.

RISTO and SEPPO obviously differ in a number of respects. First, while ristos are identified on the basis of perception alone, seppos are determined on the basis of perception and negotiation. As a consequence, no player could identify seppos on his own. Second, in RISTO we have a fixed set of objects that fall under the concept 'risto;' 'risto' thus has a 'fixed extension.' The same does not hold for SEPPO. 'Seppo' does not have a fixed extension. After all, at any given time no more than three objects are definitely seppos. And which three objects fall under 'seppo' changes over time—as a new object is entered into the

[21] Hopefully these are not familiar words in any language with which readers are familiar; I have chosen two Finnish male names in order to make sure that more readers will associate a meaning with these terms.

[22] Although it is not important for my argument, we might imagine that *A* and *B* play competitively. The more ristos *A* identifies in a given amount of time, the more points he gets, and then the roles are reversed.

array, the eldest one is dropped from the array. Note also that no object falls under 'seppo' prior to the vote taken by A, B and C. It would be incorrect to say things like 'many objects in the room really are seppo, it just so happens that the players haven't found them yet.' To make such claim would be to overlook the essential openness of similarity judgments. It might well be the case that at one stage in the game, and relative to one array, some object is collectively judged to be a seppo; while at another stage in the game, and relative to a different array, the very same object is judged not to be a seppo. In RISTO similarity judgments are less central; indeed, we might even say that all ristos are (type-)identical insofar as they all bear the mark of the stamp. Finally, and following from the above, it makes sense in RISTO to say that a successful player makes progress towards identifying all of the ristos. But in SEPPO this is not the case. Although the array drifts over time, it does not drift towards 'the final set of all seppos.'

We can also use our games to get a feel for how truth figures in meaning finitism and meaning determinism. We can bring in truth in two ways. On the one hand, we can ask what makes statements 'object n is a risto' or 'object n is a seppo' true. On the other hand, we can inquire what happens to truth if 'risto' or 'seppo' turned out to be fancy stand-ins for the word 'true.' I shall take up the two options in turn.

What makes statements 'object n is a risto' or 'object n is a seppo' true? The answer is not hard to find. 'Object n is a risto' is true is true if and only if object n bears the mark of the stamp. This truth is 'recognition-transcendent;' it holds whether or not we or player A recognize it. Moreover, there is a specific number of true statements of the form 'object n is a risto.' And that number of statements does not change. And thus it makes sense to say that a successful A discovers more and more truths. Again, the situation with SEPPO is different. 'Object n is a seppo' is true if and only if object n is judged (by the community) to be similar to the three elements of the current seppo-array. Although the communal similarity judgment is sensitive to empirical properties of encountered objects, the truth of 'object n is a seppo' is not recognition-independent; it only holds if the players recognize it as such. Truth is determined and fixed only at a time; but one and the same object can be judged to be a seppo at one time, and, at a later time, when the array has changed, it might be judged not to be a seppo. There is no fixed number of statements with the form 'object n is a seppo' that is true over time. A successful A cannot be said to discover more and more truths. And talk of getting ever closer to the truth does not make sense.

What happens to truth if 'risto' or 'seppo' turned out to be fancy stand-ins for the word 'true?' To make sense of this scenario, we have to slightly change our games. Imagine that the two games were not played with respect to normal physical objects but with respect to sentences. Imagine these sentences written on pieces of paper and distributed over the room. In RISTO, B stamps some of the pieces of paper; in SEPPO, B and C select three sentences as 'seppo.' The consequences should be obvious. In RISTO it again makes sense to say that truth is recognition-transcendent; that 'true' has a fixed extension; and that a successful A gets closer to identifying the truth. A successful A 'tells the truth.' In SEPPO truth is not recognition-independent; 'true' does not have a fixed extension; only a statement that is similar to the array-statements is true; and a successful A does not get closer to identifying the truth. A successful A does not 'tell the truth.' There simply is no truth there waiting to be told!

Needless to say, these games are remote from theories about meaning and reality that have been proposed by philosophers.[23] And yet, I do feel that they capture the gist of meaning

[23] In order to bring our two games closer to real life (i.e. real theories), we would of course have to make a whole number of changes to them. Here are the most important ones. In

finitism and other theories of meaning. Meaning finitism is committed to thinking about meaning on the model of SEPPO, other theories of meaning tend to commit themselves to construing meaning on the model of RISTO. Who then is right? I have tried to answer this question—in favour of meaning finitism—in the main text.

---

both games, we would have to allow for many classificatory words, rather than just one. In RISTO, there would be no turn-taking amongst the players; player A would be science; and player B (the "stamper") would be God or evolution. In SEPPO, we would have to drop the stipulation according to which the array can never have more than three elements. We also should allow that the elements of the array are "exemplary cases," or "paradigms" rather than just any classified elements. And we have to change the rules for dropping elements out of the array: which elements are dropped is itself something that must be negotiated amongst the players.

Chapter 5

SUBJECTIVE EXPERIENCE AND EXTERNAL WORLD[1]

Gaps and Bridges

Manuel Liz
University of La Laguna

There is an important epistemological gap in the road from our subjective experiences to the external world. It is very difficult to understand how it is possible to get knowledge of an external world, especially if we accept that all our experiences can be conceived as remaining the same through drastic changes in that external world. Moreover, according to some radical versions of skepticism we cannot even know whether there is such an external world. Also, there is an important, let us say, ontological gap in the road from the external world to our subjective experiences. We do not understand very well how it is possible for some parts of the external world —our bodies, brains, etc.— to bring about subjective experiences, especially if we accept that the external world can be conceived as remaining the same through drastic changes in our subjective experience. Moreover, according to eliminativism it is perfectly possible that subjective experience does not exist at all. The epistemological gap and the ontological gap have puzzled

[1] This work has been supported by Research Project BFF2000-0082 ("Programa Nacional de Promoción General del Conocimiento", Spain) and by Research Project PI 2001/068 ("Consejería de Educación del Gobierno Autónomo de Canarias"). I wish to thank Moisés Barroso, Rafael Herrera, David Pérez Chico, María Ponte y Margarita Vázquez —all of them members of those projects— for their insightful discussions. Also, I would like to thank Jesús Ezquerro and Marcelo Sabatés for their valuable comments in the presentation of this work at the *Seventh International Colloquium on Cognitive Science* (Donostia, 2001). In that context, I am in particular debt with Timothy Williamson for some useful suggestions.

K. Korta and J. M. Larrazabal (eds.), Truth, Rationality, Cognition, and Music, 85-108.

many philosophers.[2] My purpose here is to explore some connections between both gaps in the hope that an analysis of their relationships can give us some clues to bridge them.

In order to analyze what I am calling the epistemological gap and the ontological gap, it will be very important to make clear what is involved in them. This will be done in sections 1, 2 and 3. Then, in section 4, I will emphasize a quite simple but crucial dependence between both gaps. In a nutshell, to take seriously the ontological gap entails that we have been able to bridge, in some way or other, the epistemological gap. At this point, there would be two possible ways to explain how, in that case, we are in fact able to be in epistemic touch with the external world. Representationalism offers one of those possible ways. The other way is offered by some kind of direct realism. Representationalism is the more common option. However, the epistemological gap is supported by certain possibility of conceiving changes in the external world without any change in our subjective experiences. And representationalism never has had the enough modal force to exclude that possibility of conceiving. In sections 5, 6 and 7, I explore the other option, a kind of direct realism. With the aim of providing a strategy to limit the conceivability involved in the epistemological gap, I will introduce some concepts such as those of "knowing a sample," "knowing through samples," "indexical actualization of dispositional properties" and "typical subjective experiences." To take seriously the ontological gap would entail that we be able to know samples of typical subjective experiences actualizing certain dispositional properties of the external world, not being possible to conceive changes in the actualizations of those dispositions without changes in our subjective experiences. This would offer a very strong non representationalist explanation of how we are able to bridge the epistemological gap assuming that we take seriously the ontological gap. Finally, in sections 8 and 9, I will assess the extent to which the same explanatory strategy could be successfully applied to the ontological gap itself.

[2] For recent work concerning the epistemological gap, see McDowell (1994), Stroud (1984), Stroud (2000), and Williams (1978). With respect to the ontological gap, see Chalmers (1996), Jackson (1983), Levine (1983), Levine (1993), McGinn (1991), Nagel (1979), and Tye (1996). For some connections among both gaps, quite close to the approach that will be defended in the paper, see Putnam (1994 and 1999).

1. THE CONCEPT OF SUBJECTIVE EXPERIENCE AND THE CONCEPT OF EXTERNAL WORLD

Let us begin making clear the concept of subjective experience and the concept of external world.

- By "*subjective experience*" I understand mental states with a qualitative content of which we are conscious, being the sort of consciousness relevant here, what sometimes is called "phenomenal consciousness" or the consciousness from "the first person point of view".
- By "*external world*" I understand everything that exists but we do not reflectively recognize as being part of our subjective experiences. The external world is sometimes also called "the objective world" or the world from "the third person point of view".

Some points must be noted in relation to the above characterizations. The first one concerns the concept of external world we have just introduced. That concept is very *generic*. Beyond the claim that there exists an external world, there is no qualification indicating the peculiar ways the external world can be. However it is possible to elaborate other concepts of external world which are not so generic. In fact, we can have many sorts of propositional attitudes *about* the external world, i. e., propositional attitudes about what exists beyond what we can find reflectively as being part of our subjective experience. For instance, we can have beliefs about the *nature* of the external world. This would offer more *enriched* concepts of the external world. Moreover, as we know, these more enriched concepts of the external world can be of an indefinite variety. We can conceive the external world in a narrow physicalistic way, or we can conceive it as containing colours, sounds, and other so called "secondary qualities," or we can conceive the external world, as Berkeley did, as containing only souls and a God, etc. These would be very different enriched concepts of the external world.

I would like to note two other points. One of them has to do with the contents of subjective experiences. In principle, we could have subjective experiences *of* the external world. Our concept of external world does not exclude that possibility. That the external world is what exists beyond what we can find reflectively as being part of our experiences does not entail our not having experiences *of* that external world. To have experiences *of* the external world would be simply to have experiences *of* what exists beyond what we can find reflectively as being part of our subjective experiences.[3]

[3] This is not a mere verbal trick, but a definitory feature of intentionality. In exactly the same sense, we can say, for example, that a book about New York is a book which says certain things *of* New York without being New York, in any interesting sense, *part of* the book.

The other point has to do with our competence to grasp and use the concept of subjective experience, the generic concept of an external world and any enriched concept of that external world. The concept of subjective experience is less demanding than the generic concept of an external world, and that generic concept is much less demanding than other, more enriched concepts. Having the concept of subjective experience requires no more than having the concept of a mental state with a qualitative content of which we are conscious. In contrast with that, having the generic concept of external world requires a certain *reflective* move towards what we can find in our subjective experiences. Moreover, to use the generic concept of external world requires that we are able to claim the existence of something beyond those experiences. Finally, to be able to claim that we have experiences of an external world such as it is conceptualized in our propositional attitudes —for instance, in our beliefs—, i.e., to use any enriched concept of external world, requires another *much more reflective* move than simply to be able to claim that there is something that exists beyond what we can find reflectively as being part of our subjective experiences. It demands that it be conceptualized in certain specific ways.

A consequence of the last remarks is that even if we find good reasons to claim that we have experiences *of* the external world, we cannot directly take those subjective experiences as being experiences *of* the external world such as it is *conceptualized* in our propositional attitudes —for instance, in our beliefs—. In order to achieve this second epistemic position, other reasons would be needed.

2. TWO GAPS

As we have said, there would be two important gaps between the external world and our subjective experience. One of those gaps would have an epistemological character; the other one would be ontological. Let us define them in more precise terms:

<u>The epistemological gap</u> (hereafter, EG) would arise from the problem of knowing the external world given that

(A) it always seems possible to conceive our subjective experience as remaining the same through many kinds of changes in the external world.

EG is directly connected with the classical problem of skepticism about the external world and with problems involving "brains in a vat", etc. EG would put in difficulty our more common conceptions about the *nature* of the external world, even the very *existence* of any external world in some radical skeptical versions accepting not only thesis A but also that

(A') it always seems possible to conceive our subjective experience as remaining the same even though there does not exist any external world at all.

Being A' the case, then our concept of the external world would be in danger of becoming vacuous —a concept with a very problematic application, or with no application—. It would be possible to eliminate the external world without any change in our subjective experiences. Moreover, it is possible that, in fact, there is not any external world.

<u>The ontological gap</u> (hereafter, OG), on the other hand, would arise from the problem of understanding how the external world —specially those parts of the external world constituted by our brains and bodies— can generate subjective experiences given that

(B) it seems always possible to conceive the external world —especially our brains and bodies— as remaining the same through many kinds of changes in our subjective experience.

OG is directly related with the so-called mind-body problem, and also with problems involving "inverted qualia", "philosophical zombies", etc. OG would complicate our conceptions about the *ontological place* of subjective experience in the external world, even the very *existence* of our subjective experience in some radical eliminativist versions accepting not only thesis B but also that

(B') it always seems possible to conceive the external world —especially our brains and bodies— as remaining the same even though there does not exist any subjective experience at all.

Being B' the case, it is now our concept of subjective experience which would be in danger of becoming vacuous. It would be possible to eliminate all subjective experiences without any change in the external world —in our brains and bodies—. Moreover, it is possible that, in fact, there are no subjective experiences at all.[4]

[4] Note that in order to accept B' it is not enough to be epiphenomenalist concerning subjective experiences. It would be plausible for epiphenomenalism to accept a certain supervenience thesis about the dependence of epiphenomenical properties on non-epiphenomenical ones, in the sense that to duplicate a world with respect to its non-epiphenomenical properties entails duplicating it with respect to its epiphenomenical properties. Now, if that entailment is understood as excluding the conceivability of duplicates with respect to non-epiphenomenical properties which are not duplicates with respect to epiphenomenical ones, then it is clear that epiphenomenalism would have to reject a thesis such as B'.

3. SOME STRUCTURAL SIMILARITIES BETWEEN BOTH GAPS

There are many structural similarities between EG and OG. Let us remark on some of them.

Firstly, both gaps go against what might be called the "natural", "intuitive", "ordinary", "common-sense", etc., epistemological and ontological positions according to which the external world is, more or less, such as we experience it and our subjective experiences have a quite secure ontological place in that world.

Secondly, both gaps create difficulties for that natural position without assuming a physicalist enriched concept of the external world, i.e., without taking for granted that the external world is the *physical* world. It is very important keeping in mind the force and scope of our gaps. EG and OG pose important philosophical problems which go beyond physicalism.[5]

Thirdly, theses A and A' are *symmetrical*, respectively, with theses B and B', in relation to the kind of covariation involved. In A and A', subjective experience remains untouched through changes in the external world. In B and B' what remains untouched is the external world through changes in subjective experience.

Fourthly, the important thesis of EG is A, and the important thesis of OG is B. Thesis A' entails A, and thesis B' entails B. However, the converses do not hold true. Because of that, in order to bridge EG and OG, it is not sufficient to argue, respectively, against A' and B'. It is necessary to confront A and B, and to argue against them.

Finally, theses A and A' of EG, and theses B and B' of OG, have the *strongest modal force*. They involve *conceivability*. Theses A and A' are compatible with the existence of an external world, in any enriched sense, and theses B and B' are compatible with the existence of subjective experiences. Moreover, using recent terminology, we could say that theses A and A' are compatible with the existence of "*natural*" —or nomic, or empirical— supervenience relations of the external world on our subjective experience, and that theses B and B' are compatible with the existence of such relations holding this time from our subjective experience to the external world. There could be supervenience relations both ways involving only a certain *subclass* of possible worlds —those possible worlds sharing

---

Hence, in order to accept B', it would be necessary to also maintain some sort of "modal" eliminativism about subjective experiences. Even if subjective experiences exist, it has to be possible for them to cease existing without any other change in the world.

[5] As it will become clear in the following sections, the strategy proposed in order to bridge both gaps would also go beyond physicalism.

the natural properties of our world, or being compatible with the laws of our world, or satisfying all the empirical regularities of our world—. However, the truth of A, A', B and B' would be incompatible with the existence of "*logical*" —or conceptual— supervenience relations involving *all* possible worlds. The truth of A, A', B and B' requires at least a conceivable possible world satisfying what they say. But if there are "logical" supervenience relations of the external world on our subjective experience, then there cannot be any such world with respect to A and A'. And if there are "logical" supervenience relations of our subjective experience on the external world, then there cannot be any such world with respect to B and B'.[6]

4. A VERY SIMPLE, BUT CRUCIAL, DEPENDENCE

In any case there are not only similarities between EG and OG. No less important than the similarities above mentioned is the following kind of *dependence*:

– To take OG seriously entails our being able to bridge, in some way or another, EG.

To take OG seriously would be to take it as creating a genuine problem about the place of subjective experiences in the external world. This *presupposes* that in fact we have been able to know that there is an external world, and it *presupposes* that in fact we have been able to know something about that external world. OG only makes sense from that epistemic background.

Let us consider, for instance, the mind-body problem —the most important problem linked to OG—. If we did not presuppose that there is an external world, which is not part of our subjective experiences, and if we did

[6] We can assume for simplicity the concept of "strong supervenience" proposed by J. Kim, or may be his concept of "strong supervenience for coordinated multiple domains". See Kim (1993). However, nothing in our analysis will depend on that assumption. In contrast, the distinction between "natural" supervenience relations and "logical" ones would be very important in what follows. Also, it is very important in the approaches of Chalmers (1996) and Jackson (1982). Chalmers (1996:34-38) offers the following explanation of the distinction: "B-properties supervene *logically* on A-properties if no two *logically possible* situations are identical with respect to their A-properties but distinct with respect to their B-properties. (...) Natural supervenience holds when, among all naturally possible situations, those with the same distribution of A-properties have the same distribution of B-properties: that is, when the A-facts about a situation *naturally necessitate* the B-facts. (...) This co-ocurrence need not hold in every logically possible situation, but it must hold in every naturally possible situation. (...) It is clear that logical supervenience implies natural supervenience. (...) The reverse does not hold, however (...)"

not presuppose that we know a number of things about the nature of that external world, then we would not have a genuine mind-body problem. We would have only a, let us say, "imaginary" mind-body problem, not a genuine one. In order to have a genuine mind-body problem, we have to know many things about the composition, structure and dynamics of our bodies and brains as entities of the external world, and we have to wonder how our subjective experiences can be generated from all of that. OG only makes full sense if knowledge of that kind is taken as given.

Being serious about OG, and taking it as creating a genuine problem, entails that in some way or other *we have been able to bridge EG* and, hence, that we have been able to reject theses A and A'. This kind of dependence is quite simple and obvious. Up to a certain extent, it is also a *trivial* fact. However, that fact will have a crucial importance in what follows.

5. HOW ARE WE ABLE TO BRIDGE EG ASSUMING THAT WE TAKE OG AS CREATING A GENUINE PROBLEM?

Now, I want to pose the following question: How are we able to bridge EG assuming that we take OG as creating a genuine problem?

The situation is the following one. We are serious about OG. We take it as creating a genuine problem. Hence, we have been able to bridge EG. Hence, we have been able to reject both A and A'. The problem is: *How have we been able to do it?*

It is important to stress once again that we are assuming that we take OG seriously. It is only through that supposition that we can try to explain how we are able to reject theses A and A' of EG. First, we will discuss how we are able to reject thesis A'. Then, we will discuss how we are able to reject thesis A. Let us begin with thesis A'.

How are we able to reject A'? It is clear that if our *generic* concept of external world could not be a vacuous concept when we take OG seriously, then thesis A' could be rejected. Now, can our generic concept of an external world be a vacuous concept in that situation? The answer is negative. If we take OG seriously, then we are presupposing that our generic concept of the external world *cannot* be vacuous. If we take OG seriously, then we are assuming that an external world exists. But not only that: because we are worried about how to understand our subjective experience in that external world, we are assuming that our subjective experience *would not be the same* if there were not such an external world.

Given that we are serious about OG, there would be no need to appeal to any sort of sophisticated transcendental argument[7] in order to reject thesis A'. It would be enough to note that when we take OG as creating a genuine problem, we are assuming that there *must* be something beyond what we can find reflectively as being part of our subjective experiences.

That way, we could buy a cheap "*logical*" supervenience of the external world, understood in a generic sense, on our subjective experience. We could not conceive changes in the external world, understood in that generic sense, without changes with respect to the existence or non-existence of our subjective experience as a whole. If the external world in that generic sense were to change, then our subjective experience would cease existing too. However, the generic concept of an external world is only a concept referring to what exists beyond what we can find reflectively as being part of our subjective experiences. So, the only variation admissible in the external world so understood would be also a variation of *existence*. The external world understood in that generic sense only can exist or it can cease existing.

In other words, the non vacuous character of the generic concept of external world involved in our acceptance of OG would make us able to reject A'. But it would offer no help in relation to thesis A. In order to reject A, we need to make use of other *more enriched concepts* of external world.

Now, it is time to ask *how we are able to reject thesis A*. We must recall again that our problem is not whether we can reject A. By assumption, we take OG as creating a genuine problem. Therefore, we have to be able to reject not only A' but also A. Thus, the real problem is not whether we can do it. The real problem is one of *explaining how we are able to do it*.

In order to explain how we can reject A, there would be two open options:

1. representationalism, or
2. somc kind of direct realism.

Certainly, representationalism is the more common option in order to bridge EG. It appeals to the existence of representations playing the role of unavoidable *intermediaries* between the external world and our subjective experiences. However, no representationalist position can achieve the modal force able to exclude A. It is always possible to *conceive* variations in the nature of the external world without any variation in our mental representations. By themselves, representations are not able to maintain any "logical" relation with what they represent.[8]

[7] Transcendental arguments as those that are analyzed and criticized by Stroud (1984).

[8] Locke is one of the classical sources for representationalism. The mind would always work on representations or "ideas" of the things that we perceive or think about. As we know, in recent cognitive science the analogy between minds and computers has promoted a serious revival of representationalism. Both minds —or brains— and computers manipulate

Then, let us explore another epistemological position close to what is traditionally known as *direct realism.*[9] Our main target is to look for certain epistemic situations in which we could achieve a "direct", non-representational connection between our subjective experiences and the external world, understood this time in a certain enriched sense, such that we *cannot conceive* changes in that enriched external world without changes in our subjective experiences.

The possibility of being in such epistemic situations entails that the conceivability present in A can sometimes be only an apparent conceivability: *we would not really conceive what it seems is conceived.* From the perspective offered by those epistemic situations, we have to be able to find an enriched concept of external world with enough modal force to go against the modal force of A.

6. INTRODUCING NEW CONCEPTS

I am trying to offer an explanation close to direct realism of how is it possible to reject thesis A, and so, of how we are able to bridge EG, given that we take OG seriously. First, however, it is necessary to introduce some new concepts. Let us begin with the notions of knowing a sample and knowing through samples:

We know a sample of an x being F $=_{\text{Def.}}$

---

symbols representing aspects of the world. See, for instance, Fodor (1987: p. 17). The crucial problem for both old and new versions of representationalism is to understand how the mind can escape from its world of representations and how representations can manage to acquire a content pointing beyond themselves. This was exactly the objection Berkeley posed to Locke. It is supposed that "ideas" in the mind represent something else, but the mind has no access to this "something else" except by forming other ideas. It is exactly that lack of access which gives substance to thesis A, i.e. the possibility of conceiving changes in the external world without any change in our subjective experiences.

[9] The most classic defender of something we could call "direct realism" was Aristotle. In his *De Anima*, for instance, we can read that in perception the soul takes on the form of the object perceived without its matter. Here, Aristotle appears to be identifying the properties of the perceived objects with some properties of the perceiving sense organs. The perception would be direct in the sense that no mental "messengers", no unavoidable mental intermediaries, have to be perceived or sensed in order to perceive the properties of the perceived objects. Austin (1962) is another indispensable reference for direct realism. Even though Austin himself says that this position would be "no less scholastic and erroneous than its antithesis" —see the end of Lecture I in Austin (1962)—, that remark has more to do with the notion of "material thing" than with the rejection of sense data as unavoidable epistemic intermediaries. More recently, Putnam (1994) has urged taking direct realism seriously as a promising way to face the problems generated by representationalism, both in the fields of perception and cognition.

1. we are able to identify an x being F , and
2. we cannot conceive a more successful, or clear, case of an x being F.

In other words, knowing a sample of an x being F would be to know an x being F in the most perfect way we can conceive it. From this notion of knowing a sample, we can define another related notion.

We know through samples, or we know by samples, that y is F $=_{\text{Def.}}$

1. we know a sample of an x being F, and
2. we know that y is relevantly similar to x.

Knowing a sample of an x being F would be a special case of knowing *through* samples that that x is F. It would be that special case in which x is identical to y. We will say that we know *merely through samples*, or that we know merely by samples, in all the other cases where which x is not identical to y.

Our epistemic life is really full of knowledge through samples. Some simple instances are the following ones: knowing that something is a token of a certain word or sentence, knowing that something is the flag of the Basque Country, knowing that some object in the sky is an airplane, knowing that something has the same white colour as my shirt, etc. In all of these cases, our knowledge through samples entails 1) that we know samples of the word or sentence in question, that we know samples of the flag of the Basque Country, that we know samples of airplanes, that we know samples of things having a white colour, etc., and 2) that we know that the former things we claim to know are relevantly similar to those things of which some samples are known.

We can be wrong when we claim that we know merely through samples that something is F. Also, we can be wrong when we claim that we know a sample of an x being F. However, if we really know a sample of an x being F, then we cannot be wrong and we cannot conceive that we can be wrong. Moreover, the —of course, fallible— test to determine if we really know a sample of x being F is to try to conceive more successful cases of an x being F. If we can do it, then our knowledge was not knowledge of a sample of an x being F.

Another important notion I want to introduce is the following one:

A set of properties $G=\{Gi,Gj,...,Gk\}$ *actualizes indexically* a set of dispositional properties $H=\{Hi, Hj, ..., Hk\}$ in certain particular circumstances C $=_{\text{Def.}}$

1. in circumstances C, some properties belonging to G are exemplified,
2. $H'=\{H'i, H'j, ..., H'k\}$ is the set of non-dispositional properties associated to dispositional properties H as the properties that, being exemplified, would constitute their actualizations, and

3. *H´* logically supervenes on *G* in the circumstances *C*.[10]

The non-dispositional properties *H*' are the properties that would constitute the actualizations of properties *H*. They would be exemplified when the dispositions *H* are performed. Let us call them the "*actualizers*" of *H*. For instance, the dispositional property "to be able to be dissolved in water" would have the property of "being dissolved in water" as its actualizer, the dispositional property "to be able to have a red colour" would have the property "having a red colour" as its actualizer, and so on. To exemplify one of the properties *H*' would be to actualize one of the dispositions *H*. This is the conception of actualizing a dispositional property we are adopting. Dispositional properties have other non-dispositional properties as their actualizers, and to actualize a dispositional property is to exemplify its associated actualizer.

We are assuming that dispositional properties can be exemplified without being actualized, i.e. without being exemplified the associated non-dispositional properties which are their actualizers. It is possible, for instance, that something is able to be dissolved in water without being never dissolved in water, and that something is able to have a red colour without ever having a red colour. Also, we will assume that to exemplify any non-dispositional property able to be the actualizer of a dispositional property entails exemplifying that dispositional property. For instance, being dissolved in water entails to be able to be dissolved in water, to have a red colour entails to be able to have that colour, etc.

The important idea underlying the notion of indexical actualization is that, in certain circumstances *C*, some properties *G* get to have a "*logical*" connection with the set of actualizers *H*' of some dispositional properties *H*, being supervenience relations the way in which such "logical" connection is articulated. To exemplify properties *G* in the circumstances *C* satisfies a *logically sufficient condition* for the exemplification of some of the non-dispositional correlates *H'* actualizing some of the properties *H*. That connection would be "logical" in the sense that there is no possible world we could *conceive* in which the circumstances *C* hold, and the condition of exemplifying *G* in *C* is satisfied, but the non-dispositional correlates *H'* actualizing the relevant dispositional properties *H* are not exemplified.

[10] Again, we can assume here Kim's concept of strong supervenience and the distinction between "logical" and "empirical" supervenience above mentioned.

7. COMING BACK TO THE REJECTION OF THESIS A

With the help of the concepts just introduced, we can propose now an *explanation* of how are we able to reject A, and hence EG, given that we have taken OG seriously. The explanation would be constituted by the following steps:

1. First, we define the notion of typical subjective experiences:

   A certain subjective experience would be typical $=_{\text{Def.}}$ with respect to the circumstances in which the experience takes place, our reflection cannot but suppose a relation of *indexical actualization* between certain properties of the experience (for instance, "having a visual experience including a red colour in its qualitative content") and certain dispositional properties of the external world (for instance, "to be able to have a red colour").[11]

2. Now, the explanation would state that if we take OG seriously, then *we have to know samples of typical subjective experiences.*

   If we take OG seriously, then we have to be able to identify typical subjective experiences. We have to be able to identify circumstances in which our subjective experiences indexically actualize certain dispositional properties of the external world. There would be dispositional properties in the external world such that their actualizations in those circumstances "logically" supervene on certain subjective properties.

   Moreover, if we take OG seriously, then not only do we have to be able to identify typical subjective experiences but we have to be able to identify them without being able to conceive more successful, or clearer, cases of them. In other words, we have to know samples of subjective experiences.

[11] The concept of a typical subjective experience is the —we could say, Kantian— concept of a "*pure receptivity*" with respect to a dispositional part of the external world in certain circumstances. That state would be an epistemic state completely opposite to the Cartesian state of "doubt", and it would be also opposite to the Husserlian state of an "epojé". Up to a point, it would be similar to the epistemic state of Moore in front of his two hands. And it would be quite similar to the epistemic states described by Austin when, for instance, looking at an apple, and everything going right in our eyes, and the light being normal, and being placed neither too close nor too far from the apple, etc., we see it with a red colour. Note, however, that the enriched concept of the external world involved in a typical subjective experience would only be dispositional.

3. But, to know samples of typical subjective experiences entails that *we cannot conceive any failure in the "logical" supervenience* of certain non-dispositional properties of the external world (properties such as, for instance, "having a red colour") on certain properties of our subjective experience (properties like, for instance, "having a visual experience including a red colour in its qualitative content"), where the former kind of properties would actualize certain dispositional properties of that external world.

   In other words, to know samples of typical subjective experiences entails that having those experiences in the circumstances satisfies a *"logically" sufficient condition* for the exemplification of certain non-dispositional properties of the external world actualizing certain dispositional properties.

   This is what offers a full explanation of how we are able to reject thesis A, and therefore of how we can reject EG, given that we take OG seriously. To take OG seriously entails our knowing samples of typical subjective experiences. We know samples in which the actualizations of some dispositional properties of the external world in certain circumstances supervene on our subjective experiences. The fact that we have here a knowledge of samples entails that we cannot conceive more successful, or clearer, cases satisfying that description. And the fact that in those samples our subjective experiences are actualizing indexically certain dispositional properties of the external world entails our not conceiving any failure in the "logical" supervenience of the actualizers of those dispositional properties on our subjective experiences. In a nutshell, we cannot conceive more clear cases of an epistemic situation in which we cannot conceive failures in that "logical" supervenience. This double move is what makes possible it to resist the modal force of A and, hence, a way to bridge EG.

4. Now, our explanation could continue claiming that, with respect to many subjective experiences, we can *know through samples* whether they are typical or not. We would know whether those subjective experiences are or are not relevantly similar to some of the cases in which we know samples of typical subjective experiences.[12]

[12] That way, we could distinguish genuine claims of being experiencing the external world from cases in which we could say that we are merely dreaming, imagining, having a illusion, etc.

In the circumstances involved in our knowing samples of typical subjective experiences, we cannot really *conceive* the variations involved in A. We cannot really conceive those variations with respect to *an enriched concept of the external world as consisting in a set of "dispositions", "potentialities" or "powers"* which can be indexically actualized by our subjective experiences. That enriched concept of an external world would make us able to reject A, and to bridge EG, given that we have a genuine problem with OG.

So, there are circumstances in which we can not conceive that our subjective experiences are not —"direct"— experiences *of* the external world, understood in a special way. A very important consequence of our explanation would be that, in some circumstances, the conceivability present in A has only to be an *apparent conceivability*. In what circumstances would this happen? At least, in all those circumstances in which we are facing a genuine problem concerning OG.

Let us emphasize the last point. The modal force that makes OG a serious and genuine problem in certain circumstances is transmitted to the samples of typical subjective experiences we can know in those circumstances. In consequence, all conceivability involved in the thesis A of EG has to become an apparent conceivability when we try to formulate EG in those circumstances!

The explanation given of how to reject thesis A is really very close to direct realism. Moreover, the motivation to introduce dispositions as the basic materials of our enriched concept of an external world is closely related with two well known objections to direct realism: *the objection of "naïveté"* and *Russell's objection.*[13] On the one hand, it is easy to think of direct realism as being committed to the view that colours, sounds, etc., as perceived, are independently existing properties of objects, that the way things look or seem is exactly the way things are, even in the absence of perceivers to whom they appear that way. Direct realism is interpreted here as a "naïve" realism. On the other hand, what we have called "Russell's objection" is the problem of choosing the real colour, sound, etc., of an object given that, from different perspectives and background conditions, we always perceive different colours, sounds, etc., or different nuances of a certain colour, sound, etc.

Our approach takes typical subjective experiences as actualizing indexically certain dispositional properties of the external world. That way, it can deal with both objections. The first objection could be blocked saying that the dispositional properties are not *the same properties* than the properties that would actualize them. We would not be committed to say that

[13] See Russell (1912).

colours, sounds, etc, as perceived are independent properties of objects in the external world. The relevant dispositional properties in the external world — for instance, "to be able to have a red colour"— could exist without having any actualization. But the actualization of those dispositional properties logically supervenes on subjective experiences —experiences such as "having a visual experience including a red colour in its qualitative content"—. Hence, in the absence of perceivers there would be no colour, sound, etc., such as they are perceived.

The second objection, Russell's objection, could be blocked maintaining that even though, as we have assumed, every actualized disposition entails a dispositional property in the object, there is no need to assume that every *different* actualized disposition, entails a *different* dispositional property. With respect to many dispositional properties, exactly the same dispositional property is able to have different actualizations in different circumstances of actualization. An example of this would be the case of an object with the dispositional property of "being able to have a red colour" appearing as having a black colour in the distance. Taking into consideration the distance, "having a black colour" could perfectly be an actualizer of the dispositional property "being able to have a red colour". Our definition of indexical actualization tried to do justice to that intuition. The nuances in colours, sounds, etc., also would offer good examples of the importance of the circumstances of actualization. Properties like "having a metallic blue colour" and "having the blue colour of the sea" could be different actualizers of the dispositional property of "being able to have a blue colour" simply because the circumstances of actualization are different.[14] Hence, we could have different colours, sounds, etc., in different circumstances, and we could have different nuances of a certain colour, sound, etc., in different circumstances. Moreover, all of them could be assumed as being real. We would not need to choose because there is not any exclusion among them.

There is another important feature of our approach that must be noted. The fact that a subject knows samples of typical subjective experiences would have to be compatible with the fact that when the subject has one of those typical experiences, for instance, a visual experience including a red colour in its qualitative content, in the same circumstances and faced with the same objects, another subject can have typical experiences including, for instance, a green colour in its respective qualitative content. It is really very difficult to discard the intuition that some sort of "systematic inverted

[14] It may be convenient to note that our problem is not whether different properties can, let us say, "realize" a certain property which is the actualizer of a given dispositional property. Our problem is whether a dispositional property can have different actualizers, where the actualizer of a dispositional property is the property that would constitute the actualization of the disposition.

with that possibility. Is our direct realist approach able to make room for the existence of cases in which the qualitative fields of a subject, or of a number of subjects, are systematically inverted with respect to the qualitative fields of others? The answer is affirmative, and it will be important to see why.

Remember that in typical subjective experiences, there would be a relation of indexical actualization between certain properties of the experience and certain dispositional properties of the external world. This means that these dispositional properties would be associated with certain non-dispositional properties that, in the case of being exemplified, would actualize them, and that a certain set of these non-dispositional properties would logically supervene on a certain set of properties of our experience. That way, typical subjective experiences would achieve a logical connection with the actualizations of certain dispositional properties of the external world. The crucial point is that the relevant supervenience relation between, on the one hand, the set of properties that would actualize certain dispositional properties of the external world, the set of actualizers, and, on the other hand, the set of properties of our experience could be maintained in many ways. With respect to some subjects, circumstances and objects, the property of "having a red colour" could be related with the property of "having a subjective experience with a red colour in its qualitative content", whereas for other subjects, in just the *same* circumstances and with respect to the *same* objects, it could be the property of "having a green colour" what is related with the property of "having a subjective experience with a green colour in its qualitative content". The first kind of subjects would be actualizing, in the external world, the dispositional property of "being able to have a red colour", whereas the second kind of subjects, in just the same circumstances and with respect to the same objects, would be actualizing the dispositional property of "being able to have a green colour". But both cases could perfectly well be cases in which the subjects know samples of typical subjective experiences in exactly the same circumstances and with respect to exactly the same objects. The only thing that matters is the preservation of a "logical" supervenience relation between a certain set of actualizers, i.e., non dispositional properties actualizing certain dispositional properties of the external world, and a certain set of properties of our subjective experience. And this can be maintained through many different ways of connecting the members of those two sets of properties.

8. COULD WE APPLY THE SAME STRATEGY TO OG?

Given that we have taken OG seriously, we have sketched a *non-representationalist* explanation of how we have been able to bridge EG. According to that explanation, we know through samples that in some cases we have, let us say, *direct experiences of the external world*, those experiences consisting of actualizing indexically certain dispositional properties of that external world. That knowledge through samples in the circumstances in which we take OG seriously offered all that was needed in order to face the modal force of A.

Now, an important open question would be the following one: *Could we apply the same strategy in order to bridge OG*? In fact, an explanation analogous to the one given with respect to EG also could be given with respect to OG. The important point is that, in such an explanation, the role of "typical subjective experiences" would be played by the correlative notion of *"typical productions of subjective experiences"*.

Let us introduce that notion in more precise terms. In typical productions of subjective experiences, there is a relation of *indexical actualization* of certain dispositional properties of our subjective experiences (for instance, "to be able to have a visual experience including a red colour in its qualitative content") by certain properties of our brains, or bodies, or bodies and environments, etc. (for instance, "having a certain neuronal activity"). Just as before, our strategy could state now that we would know *samples of typical productions of subjective experience*, and that with respect to many cases of production of subjective experiences we would know *through samples* whether they are typical or not.

However, now there is an important problem. In order to bridge EG, we were assuming that we were taking OG seriously. This was what made it possible for us to be in the epistemic situation of knowing samples of typical subjective experiences. Are there circumstances in which we are in a similar epistemic position with respect to OG itself? This is the crucial question.

In any case, if our strategy were successfully applied to OG, we could obtain very relevant consequences. For instance, we would not need any other "deeper" explanation or "deeper" theory in order to solve the mind-body problem. And we would neither need any kind of "*intermediaries*" — this time, not of a "representational" kind but of a, let us say, "causal" kind[15]— between our brains, our bodies, or our bodies and their

[15] Of course, many people would say that there is a blend of representational and causal features in both cases. From the "direct realist" perspective I am suggesting, there would

environments, etc., and the production of subjective experiences. We would not need any kind of "mechanism" having states of the external world as inputs and having subjective experiences as outputs. In other words, we will have something like an ontological *"direct realism"* with respect to the mind-body problem!

Our rejection of thesis A of EG did imply the existence of a "logical" *epistemological* covariation of the external world —in a certain enriched sense— with respect to our subjective experience. And, using the same strategy, the rejection of thesis B of OG would imply the existence of a "logical" *ontological* covariation of our subjective experience with respect to the external world —again, in a certain enriched sense—. In sum, applying to OG the same strategy designed to deal with EG, we could obtain the following picture:

1. Certain properties of our subjective experience would indexically actualize some dispositional properties of the external world.
2. Certain properties of the external world —mainly, certain properties of those parts of the external world constituted by our bodies, brains, etc.— would indexically actualize some dispositional properties of our subjective experience.
3. And certain kinds of knowledge of samples and knowledge through samples would make us able to know those things.[16]

Is it possible to go a step further? Perhaps we could say that according to 1, our subjective experience would be able *to bring about* the exemplification of the actualizers of certain dispositional properties of the external world. Also, in a similar sense, perhaps we could say that according to 2, some parts of the external world would be able *to bring about* the exemplification of the actualizers of certain dispositional properties of our subjective experience. But, how can we understand what we mean here by "to bring about the exemplification"?

It is normal to assume that the only way to bring about the exemplification of a property is causal. Also, it is common to assume a certain "Humean" conception of causation entailing that, when there are causal relations, there must be laws covering those causal relations. This opens several alternatives. The way in which our actualizers are exemplified

---

not be any need to assume those features in either of the two cases —at least, not is the usual way—.

[16] Note the strong Aristotelian character of points 1 and 2. As we have said, Aristotle is the classical reference for the epistemological position known as "direct realism". Concerning the mind-body problem, he is also a classical reference for an ontological position claiming that there is a "substantial unity" between the mind and the body of a living organism. It would not be necessary to appeal to any kind of "intermediaries" either in the case of EG or in the case of OG.

could be causal or could be non causal, and there could be covering laws for those relations of exemplification or they could exist without any covering law. It must be noted that all of those alternatives would be compatible with our approach. However, the strong circumstancial or indexical nature of the exemplifications involved in 1 and 2 would suggest that, even if we assume their causal character, they would be very *singular* causal relations. Moreover, perhaps it would be necessary to postulate here the existence of singular causal relations *without* any covering law.

Many of the remarks made in relation to the direct realist strategy applied to bridge EG also could be made in relation to OG. For instance, we can recall the answer given to the objection of "naïveté". The answer was that even though the relevant dispositional properties could exist in the external world without having any actualization, every such actualization logically supervenes on subjective experiences. Hence, there would be no point in saying that, even in the absence of perceivers, the way things look is exactly the way things are. Now, with respect to OG, we could say that even though the relevant dispositional properties of subjective experience could exist without having any actualization, every such actualization logically supervenes on certain properties of the external world —mainly, of those parts of the external world constituted by our bodies, brains, etc.—.

Now, let us consider Russell's objection against direct realism. In relation to EG, the answer given to Russell's objection was that we can have different actualizers and different actualizations of the same dispositional property because the circumstances of actualizacion are different. In relation to OG, we could say something similar. The same dispositional property of subjective experience could have different actualizers and different actualizations because the circumstances of actualization are different.

Finally, as in the case of EG, we could have, in relation to OG, many different ways of adequately connecting the actualizers of dispositional properties of our subjective experience and the properties of the external world involved in our definitions. Here, the only thing that matters would be the preservation of a "logical" supervenience relation of a certain set of non-dispositional properties actualizing certain dispositions of subjective experience, a set of actualizers, and a certain set of properties of the external world —mainly, of those parts of the external world constituted by our bodies, brains, etc—.

9. WHAT IS THE SCOPE OF OUR RESULTS?

Our results do not prove that there are no gaps at all. They only try to prove that to take both gaps seriously at the same time is *self-destructive* and that, perhaps, it would be possible to bridge both gaps *in the same way*.

What we have tried to show is that to take OG seriously entails our being able to bridge EG, and that perhaps exactly the same strategy we can find in order to explain how we have been able to bridge EG also could successfully be applied to OG.

However, we cannot forget the following difference between the ways we have applied our strategy to EG and to OG. We have rejected the *modal force* of EG with the help of certain knowledge through samples of typical subjective experiences. And we have assumed that to take OG seriously supports the claim that in fact *we have* such knowledge. Also, certain knowledge through samples of typical productions of subjective experiences would allow us to face the modal force of OG. But, in sharp contrast with the former case, we have *no independent support* here for the claim that in fact we have such knowledge. So, the following question arises: *Do we have such knowledge?*

I will not give a brief answer to the last question. Instead, let us go back for a moment, and let us ask another question: *Are there other ways to face the gaps*? The answer to this second question is affirmative. There would be four main options:

1. We can consider that the kind of solution we have offered both to EG and to OG is not able to go beyond certain very restricted cases. For instance, it could be claimed that we can have certain knowledge of samples but not any other knowledge through samples.
2. Or we can consider that the kind of solution offered to EG does not apply to OG because in the second case we completely lack the relevant knowledge of samples.
3. Or we can consider that it makes no sense to face OG because there is no solution to EG, and that some kind of skepticism about the external world is the right epistemological position.
4. Or we can take the gaps in a not so serious way.

Is (4) a tenable position? How can we take the gaps in not so serious a way? The only rational way to do it would be to maintain that EG and OG do not pose genuine problems but only, let us say, "imaginary" ones. And this would involve, more or less explicitly, *coherentism* or *naturalization* with respect to EG, and *idealism* or *eliminativism* with respect to OG.

By coherentism, we can understand the large family of epistemological positions according to which our experiences cannot receive any control from the outside, so that in epistemology we do not have to take care of the external world. Naturalization would be any project of naturalizing epistemology in a way more or less close to the one proposed by Quine. A consequence of this project would be that in epistemology we do not have to take care of subjective experiences. Idealism would be the ontological position according to which all reality is mental. Finally, eliminativism would be the ontological position according to which mental phenomena, such as they are commonly described, do not exist at all. Idealism would be the ontological correlate of coherentism, and eliminativism would be the ontological correlate of naturalization. The problem is that, in the end, all these positions would reject the very terms in which EG and OG are defined: the external world or the subjective experience. This is the price of option (4).

We have to note an important point related with eliminativism. In order to bridge EG, we have taken for granted that we were taking OG seriously. We interpreted that assumption as entailing that we have certain knowledge about our brains and bodies, and about other parts of the external world, and that we face the problem of understanding how all of that can generate our subjective experiences. From the supposition that we have this knowledge about our brains, bodies, etc., we postulated the existence of certain knowledge of samples of typical subjective experiences. In these cases, certain properties of our subjective experiences would indexically actualize certain dispositional properties of the external world through the exemplification of some of their actualizers. For instance, we could say that "having a visual experience including a red colour in its qualitative content" indexically actualizes, in the external world, the dispositional property of "being able to have a red colour" through the exemplification of the non-dispositional property of "having a red colour" as its actualizer. Here, the property of "having a red colour" is taken as being the actualizer of the dispositional property of "being able to have a red colour". This would be adequate if the knowledge included in our taking OG seriously involves properties like "having a red colour" as properties of brains, bodies and other parts of the external world. In fact, this is what we have assumed in all of our examples. Moreover, I would say that it is what we would have to assume when we take for granted the relevant knowledge in order to formulate something like the mind-body problem. However, that supposition is not necessary. It is possible to argue that properties like "having a red colour" are not properties of the external world. From that point of view, in the external world there would only be physical entities and physical properties in a narrow sense excluding properties like "having a red colour". Colours, sounds, etc., only would appear as forming part of properties of subjective experiences like "having a visual experience including a red colour in its qualitative content".

How would this new supposition affect to our strategy? The answer is that it would *not* affect it in any crucial way. Now, in order to bridge EG, the actualizers of properties like "being able to have a red colour" would be physical properties in a narrow sense. In typical subjective experiences, properties like "having a visual experience including a red colour in its qualitative content" would indexically actualize properties like "being able to have a red colour" through the exemplification of some physical properties. Properties like "having such and such physical features", in a narrow sense, would now be the actualizers of properties like "being able to have a red colour". To exemplify the first property would be to actualize the second one.[17] Nevertheless, this would not be eliminativist with respect to subjective experiences. It would be only a peculiar way of constructing a bridge over EG taking OG seriously.

What about (3)? It is also important to be clear about the price of this option. It follows from (3) that, against all appearances, we never get to face a *genuine problem* concerning (OG). We never get to face, for instance, a genuine mind-body problem. And the reason would be that in no case can we be sure of having a body, or a brain, etc., as something belonging to the external world. The external world, the objective world, the world from a third point of view, etc., is always beyond our epistemic powers.[18]

Now, if we consider that the price of options (3) and (4) is excessively high to be paid, then the options that remain are (1) and (2). At this point, we must realize that those options, together with the kind of, let us say, optimistic solution we have suggested both for (EG) and (OG), would form a *continuum*. At one extreme of that continuum our solution would be placed. At the other extreme, option (2) would be placed. Between both extremes option (1) would be placed.

Taking all of this into consideration, the interesting question is not whether the solution we have imported from EG to OG is correct or not. The really interesting question is this: How can we decide which of those positions in the continuum is the more adequate one? Moreover, could

[17] From that perspective, the colours we see, the sounds we hear, etc., would not be properties of macroscopical, ordinary objects, but properties actualizing certain dispositional properties of the physical world. At the beginning of the paper we said that EG and OG pose important problems largely independent of physicalism. The main part of the paper is devoted to proposing a way to bridge EG and OG which is committed to the view that there are macroscopical objects having properties like colours and sounds. Now, we have sketched a physicalistic way to bridge EG and OG which is not committed to that view .

[18] This is the position defended by Williams (1978), with the dramatic nuance that the generic concept of the external world and some enriched concepts of it —in his terminology, both would constitute the "absolute conception of the world"— would be essential components of our notion of knowledge.

different positions in the continuum happen to be more adequate ones with respect to different particular cases of our gaps?

REFERENCES

Austin, J. (1962) *Sense and Sensibilia*, Londres, Oxford Univ. Press.

Chalmers, D. (1996) *The Conscious Mind*, Oxford, Oxford Univ. Press.

Fodor, J. (1987) *Psychosemantics*, Cambridge, The MIT Press.

Jackson, F. (1982) "Epiphenomenal Qualia", *Philosophical Quarterly*, 32: 127-36.

Kim, J. (1993) *Supervenience and Mind*, Cambridge, Cambridge Univ. Press.

Levine, J. (1983) "Materialism and Qualia: The Explanatory Gap", *Pacific Philosophical Quarterly* 64: 354-61.

---------- (1993) "On Leaving Out What It Is Like", in Davies, M., and G. Humphreys (eds.) *Consciousness*, Oxford, Blackwell, 1993.

McDowell, J. (1994) *Mind and World*, Cambridge, Harvard Univ. Press.

McGinn, C. (1991) *The Problem of Consciousness*, Oxford, Blackwell.

Nagel, T. (1979) *Mortal Questions*, Cambridge, Cambridge Univ. Press.

Putnam, H. (1994) "Sense, Nonsense, and the Senses: An Inquiry into the Powers of the Human Mind", *The Journal of Philosophy*, vol. XCI, nº 9, September: 445-517 (also in Putnam, 1999).

---------- (1999) *The threefold cord*, New York, Columbia Univ. Press.

Russell, B. (1912) *The Problems of Philosophy*, Oxford, Oxford Univ. Press.

Stroud, B. (1984) *The Significance of Philosophical Skepticism*, Oxford, Carendon Press.

---------- (2000) *The Quest for Reality. Subjectivism and the Metaphysics of colour*, Oxford, Oxford Univ. Press.

Tye, M. (1996) *Ten Problems of Consciousness*, Cambridge, MIT Press.

Williams, B. (1978) *Descartes: The Project of Pure Enquiry*, Harmondsworth, Penguin Books.

Chapter 6

THE EXPLANATORY RELEVANCE OF PSYCHOLOGICAL PROPERTIES

Marcelo Sabatés
Kansas State University

Worries about the causal efficacy of psychological properties are legion in recent philosophy of mind. Some of the sources of these worries are Davidson's anomalism, content externalism, the irreducibility of consciousness, causal-explanatory exclusion, and the structurally-based feature of functional properties. A good number of answers to these (epi)phenomenal challenges end up accepting either strongly reductive or strongly anti-naturalist positions.[1] Some of us, somewhat reluctantly, started to explore epiphenomenalism as a potential conclusion of some of these challenges.[2] In this paper I will not discuss the sources of epiphenomenalist worries. My aim here is to explore whether epiphenomenalism, and in particular the kind of epiphenomenalism that I think results from some of the sources above, can make room for the explanatoriness of psychological properties (and, perhaps more generally, of other special science properties). As it can be suspected, the answer to that question will depend on our views about explanation. Explanatory pragmatism may have an easier time making causal inefficacy and explanatory relevance compatible. My general concern is, however, whether *realist* approaches to explanation can be compatibilist in this sense. In fact, the scope of this paper is slightly more limited: I try to show that the most compelling indirect and direct arguments

[1] Kim (1998) and Hasker (1999) are examples of these opposite reactions.
[2] More or less implicitly, Jackson & Pettit (1990), Chalmers (1996), Kim (1984), and perhaps Block (1990) and Segal & Sober (1991) have accepted versions of epiphenomenalism.

K. Korta and J. M. Larrazabal (eds.), Truth, Rationality, Cognition, and Music, 109-131.

going from epiphenomenalism to lack of explanatory relevance are not conclusive. In the opening section I introduce some terminology and briefly explain some of the epiphenomenalist worries. In sections 2 and 3 I consider different versions of an indirect argument: if we want epiphenomenal properties to be explanatory within explanatory realism they have to be real, genuine properties; but this is not possible. In the section 4 I discuss two versions of a direct argument linking epiphenomenalism and causal irrelevance.[3]

1. PSYCHOLOGICAL EPIPHENOMENALISM AND ITS SOURCES

What is exactly an epiphenomenist view? First, epiphenomenalism about properties[4] of a certain kind can be partial or global. It is partial just in case it contends that properties of a certain kind are causally inefficacious *vis-à-vis* properties of some other kinds. On the other hand, it is global just in case it claims that properties of a certain kind are causally inefficacious *tout court* (including, of course, inefficacy with respect to properties of that kind). I will assume (and accept) that in the context of the several problems for mental causation, worries are about mental properties being globally epiphenomenal. Second, epiphenomenalism can be understood in a loose or in a strict sense. It is loose just in case it claims merely that properties of a certain kind are causally inefficacious. It is strict just in case it defends that properties of a certain kind (i) are causally inefficacious but (ii) are themselves causally produced. This is in fact the notion typically associated with the idea of an epiphenomenon.[5] In this context we can introduce another concept. Let's say that a theory is isolationist regarding a certain kind of properties just in case it claims that properties of that kind (i) are causally inefficacious and (ii) are not causally produced (alternatively: just in case properties of that certain kind are not involved in any causal relations whatsoever). Now, isolationism is a version of loose epiphenomenalism but it is incompatible with *strict* epiphenomenalism). I will assume that in our context most authors have strict epiphenomenalism in mind as the

[3] I am borrowing material from Sabatés (2003) for the presentation of epiphenomenalism and the ontological discussion on the reality of epiphenomena.

[4] I will frame discussions about causation and explanatoriness (and thus of epiphenomenalism) in terms of properties.

[5] I take that most traditional and contemporary *philosophical* references to epiphenomenalism are in fact references to global epiphenomenalism understood in this strict sense (see, for instance, Broad (1925), Huxley (1901), Kim (1995)). Dictionary accounts of the term are much more ambiguous.

consequence of the many problems for mental causation (but I will eventually challenge this assumption). To summarize: Mental properties are causally impotent with respect to physical or natural properties and with respect to mental properties (and, as some add, with respect to social or other higher-than-mental properties). And while in some cases there is no explicit reference to whether the mental is itself caused (leaving the result as an unspecified loose epiphenomenalism), most times it is understood that the mental is caused by the physical. In any case, what we don't find —as far as I can tell— is explicit isolationism.

Let's exemplify epiphenomenal worries with a simplified version of the exclusion problem[6] as it arises for a standard, non-eliminative naturalist. For such a view, denying the existence of souls or non-physical minds is combined with the claim that mental properties are real or genuine properties. But the naturalist also needs to affirm the primacy of physical properties over mental ones. This is achieved by the following two ideas: (a) mental properties supervene (non-causally depend) on physical properties, and (b) the physical domain is causally closed. Can mental properties be causally efficacious within this picture? The exclusion problem provides a negative answer. A mental property could be causally efficacious only if it had the potential to cause the instantiation of other properties.[7] These other properties could be mental properties, physical properties or higher-than-mental properties. Suppose that a mental property M is said to cause another mental property M*. But M* is, according to the supervenience commitment, dependent on a physical property P*, and thus P* is by itself sufficient to bring about M*. So unless we accept that mental properties are generally overdetermined we have to conclude that the only way in which M can be in some sense responsible for the occurrence of M* is by being causally responsible for the occurrence of P*.[8] Still, according to the causal closure commitment, P* must have a complete generating causal chain composed of physical properties. And it seems entirely plausible to suppose that P, the property constituting the physical base of M, is the cause of P*. But if P is sufficient for the instantiation of P*, M has no causal role in this

[6] It has been noted that the problem of exclusion is more general than the other problems, perhaps with the exception of the problem of functional properties. I have argued elsewhere (2001) that the problem of functional properties is just a form of exclusion problem.

[7] I shall simplify the presentation by eliminating in many contexts expressions like "the instantiation of" and "has the potential to".

[8] See Kim (1979) and (1989) for detailed reasons against overdetermination and other options such as partial causation.

picture unless we claim that P* is overdetermined.[9] Finally, M can be said to be the cause of a higher-than-mental property. M may be causing S*, a higher-than-mental property that depends on another mental property M*. An analogous argument shows that mental to higher-than-mental causation presupposes mental-to-mental causation, which in turn presupposes mental-to-physical causation, which in turn has been shown to be implausible. So there is no hope for this first candidate. Or M may be causing S, a higher-than-mental property that depends upon it. But this is too easy to be true. Causal relations are a kind of dependence relation, but synchronic dependence relations such as mental-physical supervenience are non-causal relations. If all we can show of a property is that it is the dependence base of a higher level one, we are not entitled to claim it is causally efficacious.[10] So there is no hope for this second candidate either. In the three cases we considered, a basic, physical property "preempted" the causal efficacy of the mental one.

Let's suppose that the problem of exclusion (or the problem of functional properties) is unsolvable and that every mental property is epiphenomenal in a still unspecified loose sense. Can we avoid the concusion that mental properties are isolated? Which properties *cause* these epiphenomenal mental properties? Obviously, other mental properties cannot do the job since, as we have seen, they are causally inert (and the same of course applies to higher-than-mental properties). Mental properties have to be caused by physical properties. But this is not possible. For suppose we say that a mental property M* (my perceiving a red object) is caused by a physical property P (the redness of the object that is on the table). Now, by the supervenience commitment, for M* to be instantiated there must be a physical base P* (an appropriate neural property) on which M* depends, and which is then sufficient for M*. So unless we admit overdetermination, P* preempts (via dependence) the causal role of P on M*. And we know, from the formulation of the exclusion problem, that overdetermination cannot be an option.

It might be replied that P has a causal role in this picture, since it is surely causing P*. However, as it was shown, M* cannot be caused by P. A property need not be a direct cause of M* (or, in general, of a given property) to be causally efficacious for it. But it needs to belong to a *causal chain* which ends up causing M*. Since P* is not causing M* we don't have the causal chain required to say that P causes M*. Now, it is more than a

[9] Strictly speaking, even if M were able to cause P* it wouldn't be *causally* responsible for M*, since the relation between P* and M* is non-causal.

[10] See below for more elaboration on this.

terminological question whether M* is caused by P*.[11] Note that if this dependence relation were admitted as a kind of causation, mental properties could be efficacious after all, since plausibly higher-than-mental properties depend on them. Also, as was noted, the formulation of the exclusion problem seems to presuppose from the outset that causation is a kind of dependence relation, but not all dependence relations involve causation. In the case under consideration the relation between M* and P* is a relation of non-causal synchronic dependence. Thus, by the same standards used in the original exclusion argument, mental properties cannot be (strict) epiphenomenal properties: Exclusion implies isolation.

Another way of illustrating how epiphenomenalism seems to collapse into isolationism is in relation with the problem of anomalism. Davidson's critics have argued that, within anomalous monism, the mental is epiphenomenal since there are no psycho-physical laws (and without laws there cannot be causal relations). But Davidson's argument for anomalous monism excludes not only psycho-physical laws (mental-to-physical laws) but also any law connecting the mental with the physical, including physical-to-mental laws. Thus, if there are no physical-to-mental laws, there cannot be —if we accept what Davidson's critics say— physical-to-mental causation. Therefore, mental properties cannot be effects and strict epiphenomenalism has to be replaced by isolationism. Our opening concerns are now more urgent: How can isolated properties be explanatory? How can they even be real?

2. THE INDIRECT PROBLEM: EXPLANATORINESS AND THE REALITY OF EPIPHENOMENAL AND ISOLATED PROPERTIES

Let us say, somewhat dogmatically for now, that given explanatory realism a property can be explanatory only if it is a *real* or *genuine* property. Explanatory realism requires "objective", "in the world" relations grounding

[11] We could appeal to authority and majority regarding this issue: With the exception of John Searle (1992), philosophers of mind defend (or assume) that a relation of supervenience or synchronic dependence such as the one between P* and M* is a non-causal relation. Some general defenses of this claim include Segal & Sober (1991), Jackson and Pettit (1990a) (they explicitly deny that a supervenience relation can be considered causal and this denial plays an important role in their formulation of the problem of functional properties), and Kim (1974), (1997). A detailed account of how causal and non-causal dependence relations differ is a difficult and neglected issue (but see Ruben (1990, chapter 7), Sosa (1980) and my (1998)).

explanations. More explicitely, the relation between the events described by explanans and explanandum has to be a real relation. And of course real, objective relations can occur only between genuine properties or events. If we accept that and add that epiphenomena are not real we have an indirect argument from epiphenomenalism to the irrelevance of psychological explanations.

So our first crucial question is: Can epiphenomenal (or isolated) properties be real or genuine properties? Those attacking epiphenomenalism from different perspectives (emergentists and reductive materialists, for instance) claim that epiphenomenalism is an unstable and even contradictory view. If the view grants the causal inefficacy of the mental it should give up the mental altogether. Eliminative or at most "preservative" irrealism[12] about the mental should follow from epiphenomenal positions. And this is even more so for isolationism. This charge is based on what Kim has called the "Alexander dictum", the idea that "to be is to have causal powers". Here are two expressions, one classical and one contemporary, of this attack:

> [Epiphenomenalism] supposes something to exist in nature which has nothing to do, no purpose to serve, a species of noblesse which depends on the works of its inferiors, but is kept for show and might as well, and undoubtedly would in time be abolished.[13]

> The total absence of behavioural effects would mean the total absence of purposes and intentions [...]. A mechanist must hold, therefore, that the principles of action have *no application to reality,* in the sense that *no one has intentions or desires or beliefs.*[14]

Ephiphenomenalists, and defenders of the explanatoriness of epiphenomenal properties, have not addressed this charge, and I want to explore if they can do it successfully. I am not claiming that answering "Alexander's objection" is *sufficient* to establish epiphenomenalism as a contender.[15] But it is at least necessary. Let us briefly clarify what is behind the "Alexander dictum".

[12] This view would reject the reality of mental properties while keeping higher-order discourse (cf. Horgan 1993).

[13] Alexander (1925), vol. 2, p. 8, quoted in Kim (1993b), p. 348.

[14] Malcolm (1968), p. 143 (my italics). Kim has recently reinforced this idea: "The reality of the mental is closely tied to the possibility of mental causation, and anyone who takes a realist attitude towards the mental must be prepared with an account of how mental causation is possible" (1991a), pp. 285–6.

[15] Answering this objection is, I think, only half of the work that needs to be done. The other half, that I am ignoring for now, is to provide a credible non-causal criterion for property reality that makes room for epiphenomenal properties but rules out "obviously gerrymandered" property-candidates.

We can agree that properties are expressed by predicative expressions (predicates for short). In a broad sense of property, every predicate expresses a property, since we can correlate each predicate with a set of actual or possible objects (or, more precisely, with a function from possible worlds to extensions). But most would agree that many properties so understood are not real or genuine properties, but arbitrary constructions sometimes called "mere-Cambridge" properties. On the other hand, we have real or genuine properties which in some sense exist in nature. We can say, following David Lewis, that among all the properties in the broad sense, most are "miscellaneous, gerrymandered, ill-demarcated. Only an elite minority are carved at the joints, so that their boundaries are established by objective sameness and difference in nature".[16] And this minority constitute the real or genuine properties, or properties in a strict sense.[17] Once we adopt a non-egalitarian view we need to say when a property-candidate is in fact a real or genuine property. A first family of criteria that might do the job expresses some kind of independence claim, usually mind-independence. This means that for an entity to be real it has to exist even if minds did not exist.[18]

There is, however, a second family of criteria that has been widely adopted, and which appeals to the causal ability of a property-candidate to enter into causal relations with other properties. The most prominent among these criteria is the one requiring causal potency or efficacy. It is in fact the application of the "Alexander dictum" for the case of properties:

(CPC) A property P is real only if it contributes to the (active) causal powers of the object that has it.[19]

In fact, (CPC) and independence principles are not necessarily competing criteria. It might well be that they supplement each other or that one reduces to the other given some assumptions.

[16] (1984), p. 227.

[17] Some would deny that a distinction can be drawn between genuine and gerrymandered (or mere-Cambridge) properties, affirming that there is no special status that the former have and the latter lacks. This view has been called the egalitarian position. I will be assuming that egalitarianism is false, and that the distinction can be drawn (cf. Hirsch (1993)).

[18] Cf., for instance, Sober (1982). He says: "Realism is a declaration of independence" and entertains criteria such as independence of human thought, linguistic conventions, our ability to discover the truth value of a proposition, etc. Cf. also Brink (1989). Mind-independence, as well as all the other criteria of this family, are expressed here as necessary conditions. This leaves open the possibility of supplementing them with other conditions (see below).

[19] Explicit defenders include, among others, Shoemaker (1980), (1979), Kim (1993b), Dennett (1984) and Fodor (1987); implicit defenders are legion.

(CPC) is a good weapon to combat undesired mere-Cambridge but it also denies reality to (strict) epiphenomenal properties and this is something that may seem too drastic. A related, more liberal causal criterion accepts properties that are impotent but are caused by other properties:

> (CRC) A property is real only if it contributes to the *active or passive* causal powers of the object that has it.[20]

3. EPISTEMOLOGICAL ARGUMENTS FOR THE CAUSAL CRITERION

The only explicit arguments for a causal criterion (and, I believe, most compelling ones) are those grounded in epistemological reasons; I will call them epistemological arguments. They share the following structure: once we deny (CPC) —or the weaker (CRC)— we face unacceptable epistemological consequences, so we are well advised to adopt a causally-based criterion. The following is an epistemological argument in which Sidney Shoemaker defends (CPC) by extracting some corollaries from its negation and trying to show that these corollaries lead us to absurd results.[21] Shoemaker says:

> If there can be properties that have no potential for contributing to the causal powers of the things that have them, then nothing could be good evidence that the overall resemblance between two things is greater than the overall resemblance between two other things.[22]

Let's suppose that two objects A and B have the same (or similar) causal powers and two other objects C and D do not. Shoemaker claims that if we deny (CPC) "[i]t might be (for all we know) that C and D share vastly more properties of the causally impotent kind than do A and B." And surely this consequence is inadmissible.

But does this consequence strictly follow from denying (CPC)? That depends, I think, on the kind of causally inefficacious properties we are willing to countenance. Recall that if we accept an isolationist but naturalist

[20] Something like (CRC) can be found in Armstrong (1979). "Passive" in this context means "being caused".

[21] This argument and the next one are in fact part of Shoemaker's larger argument for an identity criterion for properties based on identity of causal powers (cf. his 1980). According to Shoemaker, the negation of such an identity criterion would imply the negation of what I have called (CPC) (p. 214), which in turn would imply the two unacceptable consequences I will address (p. 215).

[22] (1980), p. 215.

view of the mind, mental properties are causally impotent properties which are dependent (supervenient) on causally potent (physical/natural) properties. Suppose that A and B (let's say two persons) share all (most) of their causally potent physical properties at time t. If so, they will have the same (similar) potential to cause other physical properties at subsequent times. Moreover, given the dependence (supervenience) thesis, A and B will have the same (similar) mental properties at t (the same "supervenient effects"). And further, A and B will have the same (similar) potential to contribute to other mental properties which will be dependent on A's and B's physical effects (the same potential to contribute to "indirect supervenient effects"). Thus, identity (or similarity) of causal powers is still good evidence for the overall resemblance between two things despite the fact that those things have causally inefficacious properties. Of course, these causally inefficacious properties are not randomly chosen putative properties, but properties depending on causally efficacious ones. But these are the properties we shall probably want to keep. At any rate, this seems enough to reject the generality of Shoemaker's first argument.[23]

A second and related argument offered by Shoemaker runs as follows. Suppose that there are causally inefficacious properties. Then,

> [...] if the properties and causal potentialities of a thing can vary independently of one another, then it is impossible for us to know [...] that something has retained a property over time, or that something has undergone a change with respect to the properties that underlie its causal powers.[24]

Again, the argument does not apply to all causally inefficacious properties. The reason is that the premise is not true with full generality. When we have causally inefficacious properties that do depend on causally efficacious ones, an object's properties and its causal potentialities cannot vary independently, for the causally inefficacious properties cannot differ if the causally efficacious ones do not differ. Thus, we don't have, for the case of

[23] Does multiple realizability of mental properties generate a problem for this line of reply against Shoemaker? If A and B share *all* their physical properties and so their causal powers, there is no way in which C and D can share "vastly more properties of the causally impotent kind than do A and B" (insofar as we take causally impotent properties which depend on potent ones). If A and B share *most* of their causally efficacious properties but C and D don't, it seems conceivable that C and D may share more of the multiply realizable impotent properties than A and B do. It is not clear to me what import this case may have for our *evidence* of overall resemblance of the two pairs of things. But if we are persuaded that the same property cannot be realized in wildly heterogeneous physical bases (Cf. Kim (1991b)), it is unlikely that this case be troublesome.

[24] Shoemaker (1980), p. 215.

epiphenomenal but supervenient properties, the unacceptable consequence that we cannot know whether an object has retained a property over time.

Regardless of how persuasive my replies to Shoemaker's particular arguments are, there might be a more general way of opposing them. Both arguments assume that if causal powers cannot be good evidence for the presence of a property, nothing can be. This is the main presupposition of all epistemological arguments for causally-based criteria, not just of Shoemaker's. I want to examine this assumption in relation with an argument advanced by David Armstrong as a defense of his claim that "(i)f a particular has a property, that property must endow the particular with some specific causal power." He says:

> [e]very property bestows some active and/or passive power upon the particulars which it is a property of. It must be conceded here that it seems possible to conceive of a property of a thing which bestows neither active nor passive power of any sort. But if there are any such properties, then we can have absolutely no reason to suspect their existence. For it is only in so far as properties bestow powers that they can be detected by the sensory apparatus or other mental faculty.[25]

This seems a *prima facie* strong epistemological argument for a causal account of properties, an argument which makes essential use of the notion of detectability.[26] Interestingly, the criterion that Armstrong is defending is not (CPC) but its more liberal cousin (CRC). What he requires for a property to be real is either for it to have the potential to cause other properties or the potential to be caused by other properties. This means that a property that *is caused* by other properties may be detectable regardless of whether it *causes* other properties. According to this view, non-isolated totally epiphenomenal properties would be candidates to be real properties.

Now, is Armstrong's argument a good reason to rule out isolated properties? I don't think so. If a property's being (causally) brought about may make it detectable, why couldn't a property's being generated by non-causal dependence relations make it detectable too? Suppose that Q is a property that is caused but is itself causally inert. Q will just "be there"

[25] (1979), p. 44–5. It is interesting that Armstrong concedes that causally isolated properties are conceivable before proceeding to offer the present epistemological argument —the only one he offers. This is close to conceding that there are no metaphysical arguments for (CRC).

[26] As Armstrong himself recognizes, however, there is a gap in this argument. It would need a premise like: a property-candidate that we have no reason to recognize as existent cannot be a real or genuine property. The plausibility of this premise, that I shall not challenge, probably depends on the difficult general question of whether (and to what extent) we can extract ontological conclusions from epistemological reasons.

without any possible effect. Suppose also that S is a property-candidate that is causally isolated but is supervenient or dependent upon a causally efficacious property. Why should we think that Q, but not S, can be detected by "the sensory apparatus or other mental faculty"? There are two possible reasons: Either for its "being there" or for our ability to follow the causal path which brings about Q. If we adopt the first answer, supervenient properties such as S have the same right as caused but causally impotent properties such as Q to "be there": the detectability argument cannot rule them out without begging the question. If we adopt the second answer, we lose the intuitive appeal of the argument. For the ability to follow a causal path is something that seems much more complex than —and epistemologically posterior to— the ability to detect the instantiation of a property. Moreover, it wouldn't be clear why a non-causal dependence "path" (the one generating S) is not as easy to detect as a causal path (the one generating Q) or, more to the point, as *possible* to detect as a causal path. Thus, there seems to be no justification to claim that Q is detectable while S is not. Therefore, detectability is not a good reason to prefer (CRC) over a dependence-based criterion which countenances (at least some) causally isolated properties.

Still, Armstrong's argument can be narrowed so as to defend (CPC). The key premise of this modified argument will say: a property can be detected only if it has (active) causal powers. This premise parallels Shoemaker's assumption that (active) causal powers are the only evidence for the existence of a property. Now, suppose that a property P is causally efficacious. Its merely "being there" doesn't differentiate it, regarding its detectability, from putative non-efficacious properties such as S (and for that matter, Q). Neither does P's being caused, since P shares this characteristic with Q and as we have just seen Q cannot be distinguished from S in this respect. P is detectable, and *a fortiori* recognizable as existent, because it causes other real properties, let's say P*. So we have to be sure that we can detect P* since it is P*'s being caused by P that enables us to detect P (recall that this is the only relevant difference between P and inefficacious properties regarding detectability). But we cannot detect P* unless we detect P*'s causal effects, let's say property P**. And so on. So if detectability requires causal effects, we have to go all the (causal) way in order to detect a single property! This, I think, is not something that a friend of the causal powers criterion wants to defend.[27]

[27] Note that I am not alleging a problem of ontological regress for (CPC). The objection of ontological regress would run as follows: "you require a property's having causal powers for it to be real or genuine; but this is in turn conceived as the potential to cause real or genuine properties. And this is patently circular." But see Shoemaker (1980) § VII for a reply to this objection.

So far we have interpreted the arguments for (CPC) and (CRC) in a straightforward way (let's call it the "natural interpretation"), i.e., taking it that the ability to cause *other properties in general* is the requirement for property reality. There might be, however, another way of interpreting these arguments for the causal criteria —in particular for (CPC)— which deserve attention. What if the mere fact of my ability to *perceive* a property is seen as a requirement for that property's being real? We cannot find clues for such an interpretation in Armstrong's formulation (in fact, this interpretation would undermine Armstrong's liberality towards *passive* powers), but it can be thought that this is the interpretation behind Shoemaker's arguments. But this interpretation finds more obstacles than the previous one. Perceiving something is, after all, being causally influenced by it. So the argument would be not that we have to detect a property through its causal effects on other properties, but rather that we can determine that a property is real only if we are able to perceive it. This means that those properties whose only causal powers are their capabilities of being perceived would pass the test; but this is certainly problematic.[28] Moreover, there is a specific problem for this interpretation in the context of our discussion of mental properties. Perceiving is, after all, a mental property, and as such it is suspected of being causally isolated: this is the result of the epiphenomenal arguments that generated our discussion. So, if we want to determine the reality of properties in general, we shouldn't be ready to appeal to dubious causal powers that such properties might express by "causing" a type of putative property which is itself suspected of being unable to engage in causal relations, and thus is plausibly non-real.

The "perception interpretation" of the arguments involving the notion of detection faces, therefore, serious problems. But we have previously seen that the "natural interpretation" of those arguments can also be attacked. We may conclude, then, that what seem the most promising family of arguments for causally motivated criteria of property reality are not conclusive.

It is possible to provide a different kind of argument for (CPC) based on admissible explanatory practices in science.[29] Consider the following, already classical, "egomaniacal fantasy" entertained by Fodor:

[28] First, it seems that what supporters of (CPC) have in mind is the ability of a property to be involved in different kinds of causal chains, not just in causal chains involving perception. Second, and perhaps more important, the "perception interpretation" of (CPC) goes against the spirit of the other family of criteria mentioned in section one, the independence criteria. A more detailed discussion of the perception interpretation can be found in my (2003).

[29] "Detectability" arguments, as I have said, are the only explicit arguments for (CPC). This second line of argument is inspired in (but not, strictly speaking, reconstructed from) Fodor's discussion on causal powers, explanatoriness and scientific taxonomy in his

> I define "is an H-particle at t" so that it's satisfied by a particle at t iff my dime is heads-up at t. Correspondingly, I define "is a T-particle at t" so that it's satisfied by a particle at t iff my dime is tails-up at t. By facing my dime heads-up, I now bring it about that every particle in the universe is an H-particle...thus! And then, by reversing my dime, I change every particle in the universe into a T-particle...thus![30]

We will agree that "being an H-particle" and "being a T-particle" are not real or genuine properties. And we will also agree that these "properties" are unlikely to figure in scientific taxonomies and scientific explanations. The reason can be attributed to the fact that an object's being an H-particle (or a T-particle) is wholly irrelevant to its causal powers.[31]

Along these lines we might be tempted to provide a general argument for distinguishing between real and non-real properties in terms of causal powers. It would run as follows:

(i) a property is real only if it can figure in scientific explanations (alternatively, only if it is included in a respected scientific taxonomy).

(ii) a property can figure in scientific explanations (scientific taxonomies) only if it has causal powers.

Therefore, (iii) a property is real only if it has causal powers.

In absence of a direct reason linking reality and causal powers (to prove this was the burden of the previous discussion), this argument proposes the intermediate role of scientific explanations and taxonomies to defend (CPC). Still, this indirect argument is not stronger than the previous ones.

Premise (i), at least as it is formulated, cannot be accepted. We don't want to say that the legion of macro properties that we would never find in scientific explanations are not real. "Being a bench", "having two doors" or "being a racquet" are properties that are unlikely to figure in scientific explanations. However, we are surely inclined to say that "being a bench" or "having two doors" are real properties of real objects. Similarly, these properties are unlikely to be included in the taxonomic apparatus of any science. But it seems to me reasonable to say that what counts as a taxonomically relevant property for scientific purposes does not necessarily

---

(1987) and (1991). Fodor adopts something like the "Alexander Dictum" but he does not explicitly argue for a criterion for property reality.

[30] (1987), p. 33.

[31] Fodor says: [...] what you need in order to do science is a taxonomic apparatus that distinguishes between things insofar as they have *different* causal properties, and that groups things together insofar as they have the *same* causal properties. [...] whether something is an H-(T-) particle is irrelevant to its causal powers. To put it a little more tensely, if an event e is caused by H-particle p, then the same event e is also caused by p in the nearest nomologically possible world in which p is T rather than H." (1987, p. 34).

coincide with what counts as a real property. It can be replied that a reformulation of the premise fixes this problem. For if we eliminate "scientific" from premise (i) we can avoid the counterexamples since properties like "being a bench" or "having two doors" are candidates to figure in everyday explanations (and everyday taxonomies). But this move would not satisfy many. On the one hand, if we eliminate "scientific" we lose the motivation the Fodorian argument carries. On the other hand, even the modified premise will still be unacceptable for many. For the idea that what is real *depends* on our everyday (plus scientific) explanations and taxonomies is something that a realist has to reject.[32]

We don't have to grant premise (ii) either. Premise (ii) is grounded in two assumptions that are far from obvious. How can we justify that a property having causal powers is necessary for the property to figure in scientific explanations? This answer rapidly comes to mind: First, we need to assume that having causal powers is a necessary condition for figuring in causal explanations; second, we have to accept that causal explanations are the only explanations there are.[33] The first assumption discards the possibility that causally inefficacious properties which are in some sense causally *relevant* figure in causal explanations. This is a possibility that has been suggested by some authors,[34] but I shall not insist on this. (In fact, it seems to me that non-efficacious properties might be, at most, *explanatory* relevant but never *causally* relevant.).

What I challenge is the second assumption: that there are no non-causal explanations. Fodor himself qualifies this second assumption. He says about every explanation being causal:

> No need to dogmatize, however. There may be scientific enterprises that are not —or not primarily— interested in causal explanation; natural history, for instance. And in these sciences it is perhaps not identity and difference of causal powers that provide the criterion for taxonomic identity.[35]

If some scientific enterprises allow non-causal explanations premise (ii) loses the generality that is needed to support a global causal powers criterion. As we will discuss in the next section, we have good reasons to prefer a wider, pluralistic conception of explanation in which causal

[32] In fact, this idea seems to go against the "independence-claim" that we may want to incorporate as a necessary condition for reality.

[33] Fodor seems to support these two assumptions (cf. op. cit., p. 34), but see below for a qualification.

[34] Cf. Jackson & Pettit (1990a) and Van Gulick (1993).

[35] (1987), p. 157, n. 3.

explanations are just one (though central) species of explanation.[36] Thus, if this second assumption is unfounded, premise (ii) lacks support. Therefore, neither premise of the explanatory argument looks reliable enough to support (CPC).

Still the following question can be raised: If we reject this explanatory argument, how are we to rule out, then, unwanted "properties" such as "being an H-particle"? And how are we to make sense of the idea that scientific explanations and taxonomies exclude these "properties"? The answer was elliptically suggested in my discussion about epistemological arguments. We need a criterion that is based in causally efficacious properties but doesn't rule out "well-behaved" isolated/epiphenomenal (but supervenient) properties, while ruling out "mere-Cambridge", purely stipulative properties. A candidate might look like this:

> (GCDC) A property P is real if and only if (i) P is causally efficacious, or (ii) if P is a property-candidate predicable of object x, P is involved in non-conceptual dependence relations with causally efficacious properties instantiated by x.[37]

In any case, and regardless of our views about an alternative criterion, if none of the arguments for (CPC) or (CRC) is decisive, and in particular, if none of them leads us to prefer (CPC) or (CRC) over a more comprehensive, dependence-based criterion, we don't have to accept the argument from causal inefficacy to the non-reality of the mental, and in general, from epiphenomenalism to irrealism. And this is the first step to provide a comprehensive account of how epiphenomenal/isolated properties can be explanatory.

4. THE DIRECT PROBLEM: THE CONNECTION BETWEEN CAUSAL INEFFICACY AND EXPLANATORY RELEVANCE

Making room for the reality of epiphenomena is only the first step, though. For we will encounter the following, natural reaction: once we have

[36] Cf., for instance, Achienstein (1983), Cummins (1980), Ruben (1990), Kim (1994). That explanatory pluralism can be easily accommodated within a general realism for explanations is argued in Ruben (1990) and Sabatés (1997). (This kind of pluralism *does not necessarily involve* the recognition of an autonomous category of reason-giving explanations).

[37] A more detailed (but still totally unsatisfactory) discussion of alternative criteria is provided on my (2003).

established causal inefficacy, there are at least two ways of linking it to explanatory irrelevance. One argument is quite straightforward. All we have to add is the thesis that all explanation is a causal explanation. Or, in other words, that we can explain an event or a property only if we show the cause (or at least a cause) of that event or property. The most visible form of this view defends that every explanation has to "track" an *objective* relation between the events described by the explanans and the explanandum, and this objective relation has to be *causation*.[38] If the mental is causally inert, mental terminology cannot explain.[39]

There is another argument from causal inefficacy to explanatory irrelevance. Until a few decades ago, many philosophers defended the view that rationalizing or reason-giving explanations are not causal explanations. But a highly influential paper by Donald Davidson[40] reversed this orthodoxy: he convincingly argued that the only way a rationalizing explanation can *explain* an action is if it invokes the action's cause. There are many cases in which we can have competing rationalizations for the same action, so if we do not include a causal element:

> [...] something essential has certainly been left out, for a person may have a reason for an action and perform that action, and yet this reason not be the reason why he did it. Central to the relation between a reason and an action it explains is the idea that the agent performed the action because he had the reason.[41]

If Davidson is right, we cannot explain an action if there isn't a causal relation between our mental states and it. The apparently inescapable conclusion for our purposes is that if the mental is causally impotent then intentional or rationalizing explanations are in fact irrelevant.

No matter which of the two routes we follow, mental inefficacy seems to yield explanatory irrelevance. The question here is, of course, not whether we *use* psychological explanations; it is rather whether these explanations can be *justified* by grounding them in *bona fide* objective relations. There is, however, a plausible answer to both arguments. Let's start with the Davidsonian argument. True, the only reason that can explain the action is the reason *why* the individual did it. And since "cause" is what is behind "because", the only apparent way a rationalizing explanation can explain an action is if it invokes the action's cause. If the exclusion argument is right, this has to be a neurophysiological cause of the action. But this

[38] Cf. Lewis (1986) and Salmon (1984) (Not all explanatory causalist is an explanatory realist, though. For that to be the case, realism about causation has to be added).

[39] I have attempted to address this argument in my (1997) with little success.

[40] Davidson (1963).

[41] Ibid., p. 9.

neurophysiological cause necessitates (being its supervenience base) the intentional states that are cited in the rationalizing explanation. Of course, that neurophysiological cause does not necessitate *other* potential reasons the agent might have for performing the action. So within the model we are singling out the reason *why* the agent acted as he did and discarding the other reasons *for* the action that are not related to the actual cause of the action. The fact that we are (or may be) ignorant of what is going on at the physiological level should not bother us, for the intentional states give us enough ground for believing that an appropriate neurophysiological property is doing the causal job.[42]

It could be replied that since a given mental property does not necessitate a particular realization base, unless we had knowledge of the neural or physical property that is in fact causally responsible for the action we do not have *the* reason. However, reasons are not physical or neural properties but mental ones, and all we need is to discard competing *reasons*. Thus, we can accept Davidson's premise that among competing reasons we need the one *why* the agent performed the action if we are to explain it. But this does not mean that the only way of explaining an action is by providing its cause. Language disguises the fact that it is "dependence" (causal or otherwise) that is behind "because". Once this is accepted, an appropriate path of dependence shows *why* the agent did what she did.

Let's go back now to the causalist argument. It assumes that every explanation has to "track" a causal relation. That view is far too narrow. It would even rule out, for instance, explanations of supervenient properties in terms of their supervenient bases. There are, of course, non-realist views that deny that all explanation is causal explanation, notably Achinstein's account in *The Nature of Explanation*. But since I am myself sympathetic with explanatory realism, and since explanatory realism seems to fit better with mental realism, I will focus my attention on what an explanatory *realist* has to say against causalism. But first let me briefly clarify what explanatory realism is not. An explanatory realist asks for an objective relation grounding explanations. But she does not affirm that such an objective relation is all we have to say about explanation. Explaining is an epistemological endeavor and what counts as a good explanation depends on the epistemic situation of those in need of understanding.

So what is the realist alternative to causalism? It is now widely accepted that there are metaphysically significant, "objective" dependence relations

[42] It seems clear that the problem cannot be that we are not able to give an explanation unless we are sure of which is the actual physiological cause. A bad intentional explanation would not point to a dependence path that includes the cause of the action. But, in the same way, a bad physical explanation would not point to a causal path that produces the effect.

that are non-causal. The world is structured by a variety of dependence relations of which causal dependency is just a species. Supervenience is probably the most explored among those non-causal relations but it is not the only one. Now, as it has been claimed recently, these non-causal dependence relations can be alternative objective correlates for explanations.[43] This seems wholly convincing to me. That causal explanations are just one among several kinds of dependence-based explanations is a natural view if we accept that causation is just one among several kinds of dependence relations. The specific nature of causal explanations vis-à-vis other explanations is to be found in the specific nature of causation vis-à-vis other dependence relations.[44]

A pluralist that allows explanations to track different dependence relations seems a plausible answer to the claim that epiphenomena cannot explain. However, it would be too quick to claim that every dependence relation or "path" yields a *bona fide* explanation. The main task for the epiphenomenalist then is to show that the sort of dependence relations that link, for instance, supervenient properties to the causal effects of their bases, are the right ones to ground explanations. This might seem uncontroversial for the case mentioned above: A mental property being explained by its neural base.

There is, however, a far more difficult (and crucial) case: We want to explain a behavioral property P* in terms of M, which supervenes on P*'s cause P. There is a recent attempt to provide a account for these cases. According to Jackson and Pettit the supervening psychological property "ensures" that there is a subvenient or base neural property that causes the behavior, and thus the psychological property M is in some way relevant to the behavioral property B. They say:

> The property was [...] relevant to the [effect], under a perfectly ordinary sense of relevance, though it was not efficacious. It did not do any work in the production of the [effect] —it was perfectly inert— but it had the relevance of ensuring that there would be some property there to exercise the efficacy required.[45]

So how can we turn that relevance into explanatory relevance? First, we need to define the "ensuring" (or programming) relation as the converse of a dependence relation: A property M "ensures" or "programs" a property P iff P realizes or is the supervenience base of M. Second, we need some kind of

[43] See Ruben (1990) and Kim (1994) for a defense of this view.

[44] What exactly this specific nature is raises a difficult problem. Temporal asymmetry and temporal extendibility are perhaps the best candidates.

[45] Jackson and Pettit (1990), p. 112.

explanatory principle that connects such relation with explanation. The obvious candidate seems: An ensuring property is, in appropriate contexts, a relevant explanation of the effect of its realization or supervenience base.

What should be the attitude of an explanatory pluralist regarding this model of "program" explanation? This kind of explanation is grounded by dependence relations, and in this respect there is no a priori reason to discard it (recall that a pluralist has to accept the explanatoriness of non-causal dependence relations). We cannot object that the fact that program explanations combine different relations (one causal and one non-causal). For suppose we entertain an explanation of a mental state M in terms of a cause of (its supervenience base) P, let's say P-. In such a case we explain, for instance, our current perceptual state by the existent physical condition in our environment. That physical condition P- is the cause of our current neural state P, which is the supervenience base of our perceptual state M. We have to be satisfied with this explanation if we already considered acceptable the explanations of M in terms of P and of P in terms of P-. So combination of relations shouldn't be a problem for the model.

The difficulty is that the *direction* of dependence leads from the explanans to the explanandum. For the supervenience relation seems to go the wrong way, that is from P to M. And it seems that we would need it going from M to P in order to have a (complex) dependence relation from M to P*. In fact, the ensuring relation is a "backtracking" dependence relation, and as such its explanatoriness is at risk.[46] The problem is evident from the following considerations. Does every ensuring relation make the ensuring property explanatory? The answer should be negative. The first and obvious counterexample to an ensuring property being explanatory is the case of the relation between the ensuring property and the ensured property itself. In the mind body case, the explanation-candidate "S has the neural property P because she has a sharp pain" is obviously non-explanatory. More in general, we are not prepared to accept the following:

> The parts a, b and c of a whole w have properties P, Q and R respectively because w has property T.

It seems uncontroversial that these are not respectable explanations. An implicit assumption in Jackson and Pettit's proposal is that the ensuring relation has to be accompanied by a causal relation in order to be explanatory. More generally, this means that for the ensuring property to be explanatory the chain leading from the explanans-property to the explanandum-property has to include a relationship that would be, by itself, enough to explain.

[46] This point is made in my (1997). See also Kim (1998).

There is a second situation in which the ensuring relation gives us undesired results. First, let's stipulate that my car has the property of being B iff there is a burning barn in the outskirts Donostia. Now, consider the following explanation-candidate:

> Fire engines around Donostia are working full time this evening because of my car's being B.

In this case we have an ensuring property offered as an explanations of the ensured property's effect; it seems clear that we want to reject the offered explanation. How can we rule out these kinds of explanations without undermining the explanatoriness of the ensuring relation? For my car's having property B ensures that there is a burning barn somewhere around Donostia, it does not have to be a particular fire in a particular barn of a particular place: it could be in an old wood barn or in a state of the art brick barn, it could be near the university or in Rentería (compare: a mental property M ensures that there is a relevant neural property, it does not have to be a particular neural property: it could be a neural property N or N' or N'', perhaps it could be a silicon property). And the particular fire at, for instance, a wood barn caused the fire engines to be working full-time (compare: the instantiation of the neural property N caused my walking around the library).

There is a way to rule out the undesired explanations and rule in the ones we want. The ensuring relation cannot be the result of a merely stipulative or purely attributive relation, and in general, it has to be logically or conceptually contingent that the supervenient or dependent property ensures the base property.[47] Thus, an ensuring relation is explanatory when it is tied to relations which were already explanatory and it is logically contingent.

Finally, there is a third case in which the ensuring relation needs to be fixed. Recall that causal relations are a species of (asymmetric) dependence relations; in particular they are a species of contingent dependence relations. Since the ensuring relation is the converse of an asymmetric dependence relation, the effect property ensures the cause property (the instantiation of an effect property guarantees that an efficacious property has to be instantiated although it does not guarantee that a particular property —for example the one involving such and such atoms in such configuration— has to be the case). Now, consider the following example. The coal's being hot causes the coal's being red. The coal's being hot causes the piece of tissue to smolder. So the coal's being red ensures an efficacious property, which turns out to be in a "right direction" dependence (causal) relation with the

[47] Causal and mereological supervenience relations are logically contingent while merely stipulative and "Cambridge" dependence relations are logically necessary.

smoldering of the tissue. Thus, we seem to be entitled to the following explanation:
The piece of tissue is smoldering because of the redness of the coal.
And of course we do not want to accept this as an explanation.

Once more, there is a way to exclude the undesired explanations and countenance the ones we want. In order to be explanatory, the ensuring property cannot be temporally posterior to the ensured one, and in general, it has to be synchronous with the ensured or base property.[48]

Let's now look at what the three examples we have analyzed share: they are examples of explanation-candidates that are grounded on backtracking dependence relations: ensuring explanations that are unacceptable since the arrow of dependence goes the wrong way. Do we have a way of excluding the undesired ensuring "explanations" and countenance the ones we want? The proposal has to include a set of restrictions for the explanatoriness of ensuring relations:

An ensuring relation is explanatory provided that (a) it is tied to relations that were already explanatory, (b) it is logically contingent, and (c) it is synchronic. I am not sure I can do enough here to convince the reader that this set of restrictions is not an ad-hoc set. But recall the working criterion for property reality we entertained in the previous section:

> (GCDC) A property P is real if and only if (i) P is causally efficacious, or (ii) if P is a property-candidate predicable of object x, P is involved in non-conceptual dependence relations with causally efficacious properties instantiated by x.

A criterion like this provides obvious motivation for the first two restrictions. Clause (ii) indicates that logically contingent properties supervening on causally efficacious ones are bona fide properties, and as such, the explanatory realist should be comfortable with using them (but not stipulative or ungrounded ones) for explanations. We get no help for restriction (c), though. Perhaps the only difference between the converse of contingent supervenience and the converse of causation is the way these relations enhance our understanding of an event, and that is why one is typically explanatory and the other is not vis-à-vis the effect of that event. But I won't blame you if this sound like an unacceptably pragmatic answer in the mouth of an explanatory realist.

This paper has attempted to show which should be the main concerns regarding the explanatoriness of causally inefficacious properties. I think both direct and indirect arguments can be challenged. Still, this is just a first

48 Again, recall that while causal relations involve a time-lag between their relata, supervenience relations are synchronic.

step towards a theory allowing us to *realistically* explain with *real* epiphenomena.[49]

REFERENCES

Armstrong, D. (1978), *A Theory of Universals*, Cambridge: Cambridge University Press.

Boolos, G. ed. (1990), *Meaning and Method: Essays in Honor of Hilary Putnam*, Cambridge: Cambridge University Press.

Broad, C. D. (1925), *The Mind and its Place in Nature*, London: Routledge.

Chalmers, D. (1996), *The Conscious Mind*, Oxford: Oxford University Press.

Davidson, D. (1970), "Mental Events", reprinted in *Essays on Actions and Events*, Oxford: Oxford University Press, 1980.

Fodor, J. (1987), *Psychosemantics*, Cambridge: MIT Press.

Hirsch, E. (1993), *Dividing Reality*, Oxford: Oxford University Press.

Horgan, T. (1993), "Kim on Mental Causation and Causal Exclusion", *Philosophical Perspectives*, vol. 11.

Huxley, T. (1901), *Methods and Results,* New York: Appleton.

Hasker (1999), *The Emergent Self*, Ithaca; Cornell University Press.

Jackson, F. & Pettit, P. (1990), "Program Explanation: A General Perspective", *Analysis*, vol. 50.

Jackson, F. (1995), "Essentialism, Mental Properties and Causation", *Proceedings of the Aristotelian Society*, vol. 69.

Kim, J. (1989), "Mechanism, Purpose and Explanatory Exclusion", *Philosophical Perspectives*, vol. 3.

Kim, J. (1993a), *Supervenience and Mind*, Cambridge: Cambridge University Press.

Kim, J. (1993b), "The Nonreductivist's Troubles with Mental Causation", in J. Kim (1993a).

Kim, J. (1994), "Explanatory Knowledge and Metaphysical Dependence", *Philosophical Issues*, vol. 5.

Kim, J. (1998), *Mind in a Physical World*, Cambridge: MIT Press.

Lewis, D. (1983), "New Work for a Theory of Universals", *Australasian Journal of Philosophy*, vol. 61.

Lewis, D. (1984), "Putnam's Paradox", *Australasian Journal of Philosophy*, vol. 62.

Lewis, D. (1986), *On the Plurality of Worlds*, Oxford: Oxford University Press.

Malcolm, N. (1968), "The Conceivibility of Mechanism", *Philosophical Review*, vol. 77.

McLaughlin, B. (1989), "Type Epiphenomenalism, Type Dualism and the Causal Priority of the Physical", *Philosophical Perspectives*, vol. 3.

Ruben, D. (1990), *Explaining Explanation*, London: Routledge.

Sabatés, M. (1997), "Should a Cognitive Psychologist Worry About the Causal Inefficacy of the Mental?", in Niggemeyer, B. (ed), *The Cognitive Level*, Duisburg: LAUD

Sabatés, M. (2001), "Varieties of Exclusion", *Theoria*, vol. 16.

Sabatés, M. (2002), "*Mind* in a Physical World?", *Philosophy and Phenomenological Research*, vol. 63.

Sabatés, M. (2003), "Being without Doing", *Topoi*, vol. 22.

Searle, J. (1992), *The Rediscovery of the Mind*, Cambridge: MIT Press.

[49] Thanks to audiences at the University of the Basque Country and University of Utah.

Segal, G. & E. Sober (1991), "The Causal Efficacy of Content", *Philosophical Studies*, vol. 63.

Shoemaker, S. (1979), "Identity, Properties and Causality", in S. Shoemaker, *Identity, Cause and Mind*, Cambridge: Cambridge University Press, 1984.

Shoemaker, S. (1980), "Causality and Properties" in S. Shoemaker, *Identity, Cause and Mind*, Cambridge: Cambridge University Press, 1984..

Sober, E. (1982), "Realism and Independence", *Nous*, vol. 16.

Sober, E. (1985), "A Plea for Pseudo-Processes", *Pacific Philosophical Quarterly*, vol. 66.

Chapter 7

EPISTEMOLOGY AND COGNITIVE THEORIZING

Jesús Ezquerro & Fernando Martínez-Manrique
ILCLI & Dept of Logic and Philosophy of Science, The University of the Basque Country

INTRODUCTION

The study of cognition as a natural phenomenon can affect epistemological projects in different ways. In this paper we want to examine some aspects of this influence. The paper is divided into two parts. In part one we review several ways in which epistemology can be naturalised, attending especially to the import of computational models of cognition and the different views on folk psychology. In part two we treat, as a specific case, the impact on epistemology of the putative "folk epistemology" capacity, that is, the natural ability to evaluate epistemic situations.

K. Korta and J. M. Larrazabal (eds.), Truth, Rationality, Cognition, and Music, 133-156.

PART I. WAYS OF NATURALISING EPISTEMOLOGY: THE IMPACT OF COGNITIVE THEORIZING IN EPISTEMOLOGY

1. NATURALISING EPISTEMOLOGY AND THE COGNITIVE TURN

Generally speaking, philosophical naturalism covers three different theses:[1]
a) The methodological continuity between philosophy and science.
b) The rejection of mind–body metaphysical dualism.
c) Holding externalism in epistemology.
Almost all philosophers endorse theses (a) to (c) altogether, even though there are exceptions. It is often easy to find a mistake about the term 'externalism', that can be conceptually misleading. In the tradition that goes from Kuhn and Lakatos to the Strong Program, 'externalism' is understood as the influence of external factors (say, social causes) on epistemic decisions. Some philosophers (i.e., Lakatos [1970, 1971]) allow both internal and external factors; others (i.e., the Strong Program [Barnes & Bloor, 1982; Bloor, 1981]), on the contrary, admit only the external ones. Lakatos' views are compatible with the rejection of (b), and even with the rejection of (a). For example, even if it is unclear whether Lakatos endorsed explicitly metaphysical dualism, his asymmetry thesis claiming that (in the process of reconstructing the history of science) we should look for (external) causes *only* when we are unable to find a rational explanation, could have been signed by Descartes himself. Further, this thesis seems to be closer to the rejection of (a) than to its acceptance, in so far as it involves either a view of philosophy as autonomous from science, or a view of social sciences as separated from natural ones. The latter can be also true of sociologist approaches.[2]

In the fields of philosophy of language and philosophy of mind, on the other hand, the terms 'internalism' and 'externalism' are used in a different sense, this time having to do with mental *content* individuation. This is the

[1] See Papineau (1993; 1996) for a more extended development of naturalism in epistemology involving these three theses.

[2] Fuller (1993) claims: "In terms of the recent philosophy of science debates, my position is as follows. I am a scientific realist with regard to the discourse of the social sciences. By that I mean that the best explanation for the history of all our knowledge enterprises is provided by the best social scientific theories. However, I am an antirealist about the discourse of the natural sciences, to the extent that I accept the validity of social constructivist accounts of natural scientific practices." (p. xiv)

case of Putnam [1975] or Burge [1979] (discussed at length by Fodor, 1987). Thus, you can be externalist in this sense, and nonetheless reject (a) (as Putnam does), or even reject (b) (as Kripke). Moreover, being externalist in this sense involves holding epistemological externalism, in so far as it allows epistemic evaluation to take into account factors additional to those consciously accessible to the cognizer. But the opposite need not be held: you can accept the contribution of non-conscious processes in epistemic justification without endorsing a Kripkean-Putnamian semantic view.

In a broad sense, the project of cognitive science is to investigate how information is learned, stored and used. Also in a very general sense, epistemological naturalism amounts to the view that knowledge (both common sense and scientific knowledge) is the result of causal relations between agents (people, scientists) and world (including other agents). These relations should be understood in terms of the available empirical sciences. This view is not accepted by a class of purported naturalists: sociologists claim that this is not the case for scientific knowledge [Bloor, 1991]. Instead, they hold (with Popper) that scientific knowledge is *objective* knowledge, i.e., a kind of knowledge in whose production individual cognizers (in the sense of "individual cognitive mechanisms") do not have any interesting role to play. For them, scientific knowledge is completely a social product. May be this is the reason that explains why sociologists are epistemically relativists. On the other hand, *reliabilism* is a common option for naturalists with a cognitive predilection [Goldman, 1986]: they hold that (at least) one relevant part of epistemic appraisal depends on *how* information is acquired and used. This, in turn, leans heavily on the cognitive mechanisms and constraints disclosed by cognitive science.

The interdisciplinary nature of cognitive science has yielded a great diversity of approaches within the field of epistemology. These approaches can be sorted out basically in three lines of inquiry, although it is possible to cut the cake in other ways depending on which criteria we employ. These lines are:

1. The application of Artificial Intelligence models to the analysis of the historical development of research programs.
2. The analysis of the shortcomings and biases of human reasoning capacities, revealed by experimental psychology. It is worth noting that this approach is quite different from the traditional view in epistemology that endorses and pursues categorical rationality.
3. What we will call "root consequences developers": namely, approaches that rest on whatever basic elements are posited in cognitive theorizing, and then infer large scale consequences for the conditions of possibility of knowledge, and for our second–order thinking about it, namely, epistemology. These approaches appear as the most fundamental, given

that they deal with (at least) two essential issues. The first one concerns the status of folk psychology (on which depends the overall project of traditional epistemology); the second has to do with the prospects of explanation for non-demonstrative inference, holistic or context–sensitive inference, or abductive inference. Three expressions that refer to different perspectives of the same global phenomenon. These kinds of inferences play, presumably, a crucial role in knowledge, in so far as most of our reasoning tasks make use of them.

Those three lines are not mutually exclusive. Therefore, we often find that a particular author pursues two, or even all, of them.

2. HISTORICAL ANALYSIS WITH NEW CONCEPTUAL (COGNITIVE) TOOLS

Philosophers who work on lines (1) and (2) are the new practitioners of the "historical meta-methodologies" tradition that began with Kuhn [1970]. The main difference with respect to this tradition is that historians are reproached for their "epistemological intuitionism", given that they try to assess the rationality of past epistemic decisions in the light of our current philosophical intuitions about rationality.[3] The problem for the epistemological intuitionist is that he seems unable to show that, for instance, past scientists were rational in their epistemic decisions (according to the categorical norms of rationality established by philosophy). This need not be a worry to those who, like Lakatos, are prone to apply what Popper [1972] called the "transfer of logical (rationality) to history" principle. Nonetheless, it *should* worry all philosophers that claim that we can learn our epistemic justification norms from a careful examination of the history of science (for example, Laudan [1987]). The solution of the new (cognitive) practitioners is to apply the cognitive models at hand, so as to analyse and reconstruct the history of science. This way, the result would be much more accurate and empirically based. We can include within this line authors like Bechtel, Giere, Nersessian, Solomon, and Thagard.

The focus of Giere's work [1988, 1994] has been to answer the sceptical and relativist challenges put forward by the sociologists' view of scientific knowledge. To this effect, Giere has elaborated a version of the semantic-model account of scientific theories based on the notion of *Mental Models* proposed by Johnson-Laird (1983). Nersessian [1992, 1995] and Thagard [1988, 1992], on the other hand, apply the resources of cognitive science to the analysis and reconstruction of scientific change, especially conceptual

[3] Rouse (1998) nicely describes these two lines.

change. Both share a common feature: the extensive use of *soft reasoning models*, such as analogical and similarity reasoning, as well as other ways of reasoning used in communication and presentation, such as diagrams, images, and so forth [Suchman, L. 1990]. Solomon's approach [1992; 1994; 1996] can be considered as a development of lines (1) and (2), given that she uses the experimental results of cognitive psychology in order to reconstruct historical cases. Further, she also shares with Thagard the interest on those epistemological issues related to social organisation of science and collective rationality, typical of current science.

Complementing his work on the cognitive reconstruction of history of science cases, Thagard has developed several computational models for problem solving heuristics, specifically, for abductive reasoning. He has designed some programs (PI, ECHO) that run in associative networks. In this context, he has proposed an explanation —the theory of explanatory coherence— that sets multiple constraints on the reasoning processes carried out by the associative algorithms. Using these computational tools, Thagard and his team have implemented computational simulations of some historical cases of conceptual change. In addition to his work on building computational programs, the main difference between Nersessian's and Thagard's approaches lies in that whereas the former centres her interest on the *creative* processes involved in building new concepts (context of discovery), the latter adopts a clearly *evaluative* perspective (context of justification).

Some remarks are in order. We could have worries about the philosophical relevance of this kind of approach. We could criticise it for being circular, so that it would lack interest from a philosophical perspective. This would be the case of Thagard, in so far as the results of his simulations are crucially dependent on the *input* properties (its content), which in turn depends exclusively on what the designer does at will. In other words, if we introduce in the system a set of evidence data and hypotheses, all of them adequately codified, as well as the corresponding set of constraints, the fact that the system/machine is able to do the remaining task does not matter at all. This is so because the interesting labour would be just *handmade*, say, the evidence, hypotheses, and constraints introduced. But this, as it happens in all AI simulations of problem solving, depends crucially on previous conceptual work that implies that those who are able to build a computational model to solve a problem *already* have a way of solving it *before* building the algorithm. From this perspective, our simulations would amount to a sort of mechanisation of cognitive or rational processes, but not to an explanation of them.

Thus, in simulating historical cases of problem solving, we would need a previous rational reconstruction. Once we have one at hand, all that remains is the routine, not philosophically interesting work of designing the algorithm. Hence, the traditional historian can claim that there is no *real* difference from his own approach. The reason is that the results of the simulation would depend on the rational intuitions employed in the previously made reconstruction. Moreover, computational simulation would remain *neutral* concerning the epistemic values employed during such a reconstruction: "give me a rational reconstruction of a case of problem solving and I will give you a program simulating it", could be the slogan,

We have to be cautious here, however. It often happens that machines can surprise us. It might be argued that these surprises are only apparent, not genuine, in the sense that machines are not able to create knowledge. They just limit themselves to manipulate previously introduced information. Thus, at the best, they can deploy explicitly information already contained (implicitly) within the information we previously input. Yet, even for this modest job it would be worth doing simulations. But there is more: it turns out that the situation of trying to simulate an intelligent task *forces* us to learn a lot of things about the nature of the target task, as well as about the nature and capacities of the systems capable to accomplish it. It is possible that a lot of details and constraints can remain unknown to the pure conceptual analysis, whereas they reveal themselves when we try to build a specific system capable to do the job. For this reason, the naturalist philosopher, even if she is a normativist, should not reject this type of project.

Bechtel [1995] can be included within this line. He combines a close familiarity with connectionism with an interest on the epistemological and normative issues of scientific knowledge. His main contribution is the development of heuristics and cognitive strategies —decomposition and localization— that he offers both as methodological-normative proposals and as a conceptual frame to analyse historical cases of theory change, mainly in the fields of biology and psychology. He centres his attention on the psychological constraints operating over the development of scientific theories, and his main goal is to propose dynamical models of rationality plausible from a psychological viewpoint.

So far our description of some ways in which cognitive theorizing is currently influencing epistemological issues has focused especially on the philosophical accounts of scientific knowledge (lines (1) and (2)). Nonetheless, we must note that all the aforementioned approaches involve deep, basic epistemological and ontological commitments that we have not touched yet. It is plain that these accounts use folk psychology in one way or another. They assume a view of agents (scientists) *as believers*, as beings that have *representations* and engage in reasoning processes that operate on representations.

On the other hand, the customary employment of heuristics involves the assumption that most reasoning processes are like non-demonstrative inference, in contrast with classical (logical) inference. This is a fact that no one wants to dispute, of course, but it raises a deep problem that has to do with the sheer possibility of naturalising epistemology, i.e., about the prospects that cognitive science has of illuminating our epistemological discourse.

3. COGNITIVE APPROACHES AND THE STATUS OF FOLK PSYCHOLOGY

A first question about folk psychology is whether or nor it has to be regarded as a *theory*. This is a controversial issue. The reason is that it seems a matter of conceptual commitment that if you consider **X** as a theory, then you are admitting a lot of things, e.g., as such a theory it may be true of false. To decide this, you can appeal to a number of reasons, being the most important among them whether or not its posits (its ontology: the things and properties postulated) *do refer* (really hold) within the domain. And if it turns out that the theory is false, then you should look for a replacement of such a theory. In addition, there are other constraints to be satisfied by a theory, and these, in turn, play a relevant role in deciding whether to maintain or reject it. It must have explanatory power, allow progress and integration with other theories we may have about the world, and so forth.

On the other hand, there are several features that are often attributed to folk psychology, features that we can synthesise in the following four:

a) Mental states are semantically interpretable, that is, they have *content*, in the sense that they are *about* something.
b) You can have different *attitudes* towards the same content. You can believe it, fear it, desire it, and so forth. These two things, the kind of attitude and its content, allow us to individuate the complex mental states that are the propositional attitudes.
c) Mental states play a causal/explanatory role: mental states explain behaviour *because* mental states *cause* the behaviour.
d) Mental states are *functionally discrete*. By this it is meant that they are separable entities. You can have a propositional attitude independently of your having or not others. They can be distinguished from each other.[4]

[4] This is the typical characterisation used in the literature, in order to analyse whether or not connectionist systems are able to deal with these commitments. See, for example, Horgan & Tienson (1995).

It is unclear whether psychology satisfies either the requirements for being a scientific theory, or even the four features just mentioned. Even if folk psychology were unable to satisfy almost all of them, this fact should not prevent us from regarding it as a theory, although of a special kind: a theory whose status would not be exactly like ordinary scientific theories. All this is highly dependent on your favourite basic approach to cognitive systems, together with additional considerations. There are two (and a half) influential cognitive approaches in the market: (a) classical; (b) connectionist; and, finally, the half, (c) evolutionary psychology. The reason to call it "a half" is to reconcile the following two features of the evolutionary approach: (i) its aim to constitute a full-blooded account of cognition (an alternative paradigm, if you like); and (ii) the fact that this account is not committed to a specific way of characterising the nature of psychological processes. For example, Cosmides & Tooby [1994] claim that psychological processes are computational. That's all. So it seems that this approach could be compatible both with (a) and (b).

With respect to the ontological status of folk psychology, the main controversy arises between realist and eliminativist positions, which are typically represented by Fodor [1981] and Churchland [1989], respectively. It is also noticeable that they are conspicuous defenders of the classical and connectionist approaches, respectively. But the link between classicism with realism, on one side, and connectionism with eliminativism, on the other, seems to be more a *de facto* than a *de iure* situation. Actually, the realism/eliminativism controversy began before connectionism appeared on the scene (we could trace it back to Sellars's [1963] scientific and manifest images, for example). There is also the (quasi)instrumentalist approach favoured typically by Dennett [1987], that would deny features (3) and (4) above. For the sake of space, we will concentrate on the other two positions.

As it happens, Fodor would tell us, you do not need too much effort to be a realist about folk psychology. It is, we would say, a *non-negotiable fact*. If it turns out that your favourite approach (be it classical or connectionist) gets in trouble with folk psychology, so much the worse for your approach. Thus, the task to be done is to develop a cognitive theory that maintains the basic folk psychological ontology, and whose goodness is subordinated to its capacity to abide by the explanatory standards of folk psychology. The defenders of this position, however, have argued at length that connectionism is unable to render an account faithful to folk psychology. This is a point about which eliminativists and instrumentalists do not worry, even if for different reasons. Dennett [1991] would agree with the practical indispensability of folk psychology, but he argues that we should no longer treat it as a *theory*, but as a *craft*. Being things so, folk psychology and cognitive theorizing do not compete, given that their goals are quite different. It may well be that the real underlying mechanisms responsible of behaviour have nothing to do with our folk psychological descriptions and explanations. Eliminativists, on the other hand, reply that if folk psychology does not match connectionist systems, it is folk psychology that must go (given that connectionism is a better theory of cognition than folk psychology). Nonetheless, things are not so easy to those that try to endorse together realism and connectionism.

The problem with this position is to make compatible the four typical features (1-4) of folk psychology with the putative commitments of connectionism. Horgan and Tienson [1994, 1996] summarise the main commitments in the following three:

i. Its codification is massively distributed (non-local).
ii. Hidden units are sub-symbolic: they do not have semantic interpretation.
iii. Connectionism is proposed as a full account of cognition (not only as yielding implementation models).

So, from the perspective of commitments (i)-(iii), the problem is to make them compatible with feature (4) of folk psychology, functional discreteness [cf. Ramsey et al, 1991]. Horgan and Tienson (1996) distinguish three different ways in which a psychological state type can be possessed by a cognitive system: (a) occurrently; (b) dispositionally; (c) morphologically. These three ways can distinguish kinds of functional discreteness. While folk psychology is usually committed to (a) and (b), they claim that it is also compatible with (c). In fact, they argue, folk psychology usually deals with psychological states that are both occurrent and conscious, and connectionist systems (of the sort of those proposed by Ramsey *et al.*) have no trouble to accommodate all kinds of functional discrete mental states. In fact, they argue, connectionist systems meet Marr's (1982) three-level description just as classical systems do. And if classicists do not have trouble to accommodate the higher (cognitive) level, connectionists do not either.

In general, it is typical of those connectionists that want to keep folk psychology, in contrast to eliminativist connectionists, to share a *revisionist* view. In other words, they give theoretical and ontological primacy to the algorithm and implementation levels, in such a way that the development of our theories on these levels can involve a revision of our ontological and epistemological intuitions based on folk psychology.

4. COGNITIVE THEORIZING AND EPISTEMOLOGY

To conclude this quick revision of the relations between cognitive theorizing and epistemology, we can draw several morals. In general, we can say that all the different positions are *naturalist*, in the sense that all of them share the three naturalist theses (a-c), presented in section 1, even if they do not do so in the same way. Even when they share the thesis of methodological continuity, there is an important difference between realism, on the one hand, and revisionist and pragmatic approaches, on the other. The realist + classical account brings about a sort of methodological dualism internal to cognitive theorizing: the higher level (cognitive) would be autonomous with respect to the lower levels. Classicists claim that unless we admit a cognitive architecture based on the computational theory of mind (with a set of processes that operate on symbolic-syntactically structured representations, really tokened on the algorithm level), it is impossible to account for cognitive behaviour, more particularly, to account for *rational* performance. The idea behind this is that an explanation of a rational behaviour cannot be considered as such *unless* it shows that the behaviour is caused by a process instantiating a justifying argument. This is what Cummins calls the *rationale constraint*.

Classicists claim that they have no problem concerning this constraint. Thus, the troubles fall again on connectionism. Again, if it is possible to show that classicist arguments are wrong, or if we can offer alternative ways of describing the algorithm level (e.g., in terms of dynamical systems) that account for structure sensitive processes, then connectionist systems could be capable to deal with the rationale constraint. On the other hand, we must note the inability of classicism to account for the more interesting cases of cognition, say, the ones that deal with non-demonstrative inference, abductive inference, context-sensitive inference, and so forth. In fact, Fodor [1983; 2000] explicitly sees it as a hopeless project, maintaining that computational accounts are limited to the modular systems. Yet, the claim of the connectionist camp is that neural networks seem to be better equipped to deal with this kind of inference. But the price to pay is to renounce, in some way, to the rationale constraint.[5]

[5] In this respect, it is worth mentioning a recent paper by Cummins *et al.* (2001), where the authors try to show that connectionist systems of the sort of Smolensky, Legendre and Miyata (1992) do meet the challenge of systematicity (Fodor and Pylyshyn, 1988) by means of structural encoding (in contrast to structural representations). Being able to deal with systematicity is a necessary requisite to meet the rationale constraint, in so far as reasoning involves cognitive processing on structured domains. It is too early to evaluate this proposal. Anyway, the issue is still open.

PART II. FOLK EPISTEMOLOGY AND EPISTEMIC INTUITION

1. FOLK WISDOM

Ordinary people are said to be endowed with several systems of knowledge: one that allows them to predict and explain the behaviour of others in many normal circumstances; another that allows them to predict and explain physical events in everyday contingencies; a third one that allows them to differentiate animal and vegetal species attending to their most distinctive features. These systems have received the names, respectively, of folk psychology, folk physics, and folk biology. There is much controversy about the nature and the extent of this knowledge. In the case of folk psychology, it is disputed whether it is innate or acquired (Scholl and Leslie, 1999), whether it is encapsulated or available for general processing (Gopnik and Meltzoff, 1997), whether it consists of bodies of specific knowledge or it requires specific processing mechanisms (Samuels, 1998), whether it is culturally determined or evolutionarily prearranged (Tooby and Cosmides, 1992).

There are other possible candidates for folk wisdom: folk mathematics, folk linguistics, and so on. We want to consider a variety that we can call "folk epistemology". Folk epistemology (FE) is concerned with people's knowledge about knowledge. Some questions that FE is interested in are: how people know (believe, etc) that other people know (believe, etc); in what cases people judge that a person "knows" instead of "believes"; what elements are relevant to these judgements, etc. There are two obvious relationships that come to front when one considers FE: its relation to folk psychology, and its relation to epistemology. *Prima facie*, FE seems to be a part of folk psychology. Epistemic notions such as knowing, believing, understanding, remembering, etc, are part of the set of propositional attitudes which we use in our explanations and predictions of behaviour. Hence whatever things people know about knowledge is simply a province of whatever things they know about each other's minds.

However, if we take a closer look, folk epistemology does not seem to be concerned so much with the prediction or explanation of behaviour as with a particular *categorisation*. We can compare it with folk biology. Folk biology is not concerned with, say, what the tiger will do next or why it crouched instead of running. This is a piece, if anything, of "folk tiger psychology". Folk biology is more concerned with telling a tiger from a lion, recognising different types of plants, distinguishing which plants or animals may be edible, and so on. It is a system of knowledge directed to classify different parts of the world in intelligible categories. There may be explanation and prediction in folk biology, but this is not its primary purpose. This contrasts with folk psychology, where explanation and prediction are in the foreground.

Similarly to folk biology, FE deals principally with a particular kind of categorisation, namely, epistemic categorisation. Its chief purpose is to reach a judgement about the epistemic state of an individual (who can be the judge herself). Typical questions that FE makes are: "does he know X?", "does she believe Y?", "is he justified in believing that Z?", and so on.

There are at least some natural situations for FE to operate. For instance, in a *learning situation* a teacher wants to find out if her pupil knows the lesson, if he really understands, if his erroneous answers are due to his ignorance or to some teaching failures (e.g., a bad book), etc. A second example is *cheating*, in which a potential cheater wants to find out if someone knows X, or a suspicious individual wants to find out if someone is telling a lie.

One might object that those are cases that properly belong to folk psychology, or the art of mindreading. Mindreading is all about attributing mental states to other people (Davies and Stone, 1995), so that their behaviour can be explained in terms of those mental states. However, there are two aspects to mindreading that do not have to be conflated. One of them is mind*reading* in the strict sense, that is, it is directed to find out what mental state the other person is in. The other aspect is mindreading in a *projective* sense: we use our knowledge of the other person's mental states to determine what she is going to do next or why he did what he did.

The difference between both aspects is much like the difference between theoretical and practical reasoning, distinguishable even if closely related (Harman, 1986). Sometimes we just stop short of reaching an explanation or prediction, i.e., our goal can be merely to attribute a particular mental state. For instance, in the mentioned learning situation the teacher's task ends when she finds out the state of knowledge of the pupil: there is not any further behaviour or mental state to explain or predict on the basis of that finding. At any rate, both processes, mindreading in the strict sense, and mindreading in the projective sense, can be kept separate. Now we can see that FE has to do with mindreading in the strict sense. It deals with the attribution of mental states, but not any kind of mental state, only with those of some epistemic bearing. The point is to find out whether someone really knows, merely believes, and so on. Even if we understand folk psychology in a broad sense so that FE is a province of it, FE can still be regarded as a set of distinct abilities within the complex system of capacities that constitute folk psychology.

This can be also observed in the case of folk biology. Suppose we are wondering why Peter is running scared. We look around him, see a snake, and understand his behaviour. Now, in this process we have used a piece of our folk biological knowledge, namely, that the creature over there is a snake, and that snakes are generally dangerous. Yet, we would not say that in doing this we make folk biology a part of folk psychology. Just as we do not say that any of the other cognitive capacities involved in explaining his behaviour belong to folk psychology —e.g., capacities like vision or hearing.

The second important relation of folk epistemology is with epistemology. In what sense is FE, as a mental ability or set of abilities, a part of epistemology, as a philosophical project? Traditional analytical epistemology seems to depend on FE. It relies on epistemic intuitions in order to set forth a number of principles or criteria, that constitute the epistemic notion which is the focus of the analysis. In other words, if a proposed analysis maintains that, in order to qualify as knowledge, a state has to possess such and such features, then there cannot be a paradigmatic example of knowledge that lacks those features, and there cannot be a clear example of non-knowledge that has those features. This analysis can be made for mere descriptive purposes. However, for many epistemologists the proper end is to reach a normative stage, in which we can say, for instance, that in order to attain a state of knowledge such and such conditions must be met, and such and such epistemic states and processes are the right ones.

In the following sections we will try to clarify the notion of folk epistemology, examining its putative role as a cornerstone of a number of epistemological projects, and its relation with folk psychology.

2. THE VARIABILITY OF EPISTEMIC INTUITIONS

People can judge immediately and effortless about epistemic states, both their own and alien. Part of the job of a clever epistemologist is to arrange situations to test our judgements, in order to reveal hidden features of the epistemic situation. The best cases for analysis are those that, by triggering a widespread intuitive judgement about the attribution in question, get to challenge established analyses of an epistemic notion. This is the strategy followed in the well-known Gettier examples (Gettier, 1963), and this is why they drew so much attention.

Now, suppose we have cases in which people's intuitive judgements do not agree. Then their usefulness for the epistemologist would not be so clear. Hence it might be the case that they could not support the epistemological project because the variance within epistemic intuitions is too large. For instance, Weinberg, Nichols and Stich (forth.) have argued that our intuitive judgements about when somebody knows or does not know X are subject to the influence of cultural and socioeconomical factors. In a series of experiments, they compared the epistemic intuitions of European Americans with those from people of Asian origin. Their results showed significant differences between both groups. For instance, when they used Gettier cases as the testing material, Asians tended to attribute knowledge to the person described in the case, a tendency opposite to Western subjects (and contrary to widespread intuitions in Western philosophical community). They also compared epistemic intuitions from people with low and high socioeconomical status, showing that the former were more ready to attribute knowledge in problematic cases than the latter.

Weinberg et al. use their results as an argument to attack what they call "intuition driven romanticism". This is a family of epistemological projects that take epistemic intuitions as data in order to produce normative claims about epistemic matters, such as evaluating the merits of different belief formation strategies or judging that certain doxastic structures constitute real knowledge. The fact that different sets of intuitions can lead to different normative epistemological theories should cast doubt on those normative claims. Given that there are no independent grounds to assert that Western epistemic intuitions are more granted than intuitions from Asian people, the epistemologist that relies on typical Western intuitions cannot avoid the charge of being engaged in an ethnocentrist project.

It seems that Weinberg et al's experiments are, to some extent, a case of overkilling. Interesting as their results are, the point they are hammering could have been made merely looking at divergences between Westerners in judging the different examples. In fact, these divergences may appear even within the philosophical community. Even if it is a minority that dissents from the most prevailing opinion, we cannot rule out their intuition simply by "counting heads". So the point could simply be stated thus: how is it that normative principles following from your intuitions are more sound than normative principles that derive from my intuitions, or vice versa? Are they merely differences in opinion?

To clarify these matters, we need first to examine what sort of thing is an epistemic intuition. However, there is not a univocal way to treat this notion.

3. KINDS OF INTUITIONS

Intuition is usually understood in the vague sense of immediate knowledge whose origin we are not able to state precisely (e.g., it does not come from the senses). Hence if you ask me what kind of tree this is and I just fetch for my botany book and answer "it is a pine", my statement does not count as an intuitive judgement. In an intuitive judgement one "just knows" that this is a pine and not a fir. Yet between both extremes there is a range of possible judgements that do not fall clearly into one side of the divide. What if I remember having read about it some time? What if I think somebody told me about it? When does it count as "just knowing" it? Even if it is not totally clear when something counts as an intuition, it seems to be the case that we call intuition to disparate types of judgments. We will distinguish four different kinds of cases of intuition.

Case 1: Linguistic intuitions.

We can tell without much effort which sentences are correct in our language (or our idiolect). These intuitions are not learned, neither in a formal nor in an informal setting, in the sense of receiving feedback about the well formedness of sentences. They can be refined, e.g., by taking a course in linguistics or any other method that makes one more aware about the structure of his language. But speakers of the language, even the illiterates, can agree about a vast number of well-formed sentences. Moreover, linguistic intuitions have normative force, in the sense that we can require that "you ought to say it this way, rather than that way".

Linguistic intuitions are not innate, i.e., we cannot have intuitions about language until we have acquired language. The debate about the innateness of language is irrelevant here, because even if there is something like an innate universal grammar that every new born child shares, the linguistic intuitions in question are not about this grammar. The relevant intuitions are about the grammar of the acquired language.

Linguistic intuitions are productive, in the sense that once a person has mastered a language she is able to judge about the well formedness of a (possibly infinite) number of sentences. (We say '*possibly* infinite', even if an infinite number of sentences can be recursively generated by a natural language, because there may be processing limitations that constrain the class of sentences that can be reasonably comprehended by the subject).

Case 2: Folk biological intuitions

A point can be made for the existence of an innate body of knowledge about the natural world. It may be very basic, comprising things like being able to tell an animal from a plant, to tell a human from a non-human, to tell a living being from a non-living being, to expect certain things from animals that one does not expect from non-living beings, and so on. It might not even include the ability to tell a dog from a cat. At any rate, the knowledge that it does and it does not include is something to be determined empirically.

We can locate within this class of intuitions other putative examples of folk wisdom, like folk physics or folk psychology. In this case it *is* important whether the alleged pieces of knowledge are innate or not, especially to distinguish them from the third class of intuitions that we will present. In other words, the assumption is that if there is such a thing like folk knowledge, it is hardwired. It was provided by evolution through a lengthy period of time as the individuals coped with various environmental challenges. A related assumption is that this knowledge is (more or less) fixed and universal, so that individuals going through different ontogenetical processes would end up with (more or less) the same set of folk intuitions. So it is not productive in the sense explained for linguistic intuitions.

Case 3: Biological intuitions

Most of our wisdom about the natural world is acquired through a process of enculturation. We learn facts about dogs, cats, apple trees and the like and we begin to accumulate a database that is increasingly organized in a web of implicit knowledge. Maybe "intuition" does not sound as a good label for this kind of knowledge, because we tend to regard intuitions as something more obscure, whose origin is somewhat deep and mysterious. But the chief feature of intuition is that it drives spontaneous, immediate judgements, so there is no reason why we should not treat all this culturally acquired knowledge as the basis for intuitions.

In contrast with folk biological intuitions, biological intuitions are not innate. They are not hardwired in any sense. They are learned, sometimes in formal settings, sometimes simply picking information from the interaction with conspecifics or with the world. Biological intuitions are not productive either, at least they are not so in the sense in which linguistic intuitions are. One can be happily ignorant about vast portions of biological facts, say, that X is a wolf and not a dog. However, one cannot be happily ignorant about his language. If you think that it is possible to say "the dog runs" but not "the wolf runs", then there is something deeply wrong about your mastery of English. Maybe the reason for this difference has to be found in the nature of the domain to which biological and linguistic intuitions apply, i.e., maybe our linguistic intuitions are productive in the referred sense because language is productive itself. Be it as it may, the difference with biological intuitions still remains.

Another difference with linguistic intuitions is that the latter do not arise through a process of enculturation. Even if you want to say that learning a language is a form of enculturation, there is still a notable difference. Learning a language is not a matter of acquiring a great number of facts systematically related, in contrast with the sense of enculturation that seems to be in charge of learning biology. In fact, it is reasonable to assume that most of our tacit knowledge about the world has been acquired in this way. Thus it is possible to have intuitions of this type about almost everything.

An important difference between linguistic intuitions and the other three classes is the normative force of the former in order to construct a theory about the domain in which they apply. In other words, it would be a failure to propose a theory of French grammar or a taxonomy of French sentences that took as well-formed expressions which were unacceptable by French speakers. In contrast, it would not be preposterous to propose some animal taxonomy that overtly clashed with ordinary intuitions (e.g., by classifying fungi closer to animals than to plants). Yet, there is a sense in which biology is also a normative project: it relies on an attempt to obtain the *right* classification by attending to the relevant features.

Case 4: Moral intuitions

Many people seem to have strong intuitions about what is right and what is wrong. In one sense these intuitions seem to belong to the same class as the biological intuitions presented just above. For one thing, they are also clearly obtained through a process of enculturation. There are attempts to uncover whether there are general moral rules that are endorsed universally by people from very different cultural backgrounds. Yet, even if such standards exist, here more than in any other domain variability is the norm. In fact, it is the extreme disagreement in moral judgements and its extreme resilience to change that suggests that they constitute a different category of intuitions. Barring the existence of a "moral core", moral intuitions would not be innate nor productive. They have, however, a strong normative nature, telling the person who holds the intuition what ought and ought not to be done.

4. THE NATURE OF EPISTEMIC INTUITIONS

So, what kind of intuition is a folk epistemic intuition? The answer to this question seems to constrain the uses to which epistemic intuitions may be put. To begin with, we have been conflating two different kinds of intuitions in the notion of epistemic intuition. We began by talking of folk epistemology, and then we moved to talk about epistemic intuitions. Yet, it may be the case that they do not have much in common. To put it another way, it may happen that folk epistemic intuitions are just a small subset of epistemic intuition *tout court*. Folk epistemology would be with respect to epistemic intuition much the same as folk biology is with respect to biological intuition: a nucleus or foundational cornerstone on which the rest of the intuitions can be built.

Let us consider first that epistemic intuitions are like linguistic intuitions. This appears as the most attractive hypothesis for the epistemologist, because linguistic intuitions have exactly the features she needs for her project: they have normative force and they are shared within a linguistic community. If epistemic intuition is like linguistic intuition, we can make then a comparison between variation in epistemic judgements, and variation in language. Each set of (culturally or socioeconomically influenced) epistemic intuitions would count as a "different language" and the task of the epistemologist would be to disclose the normative structure of such epistemic language.

Yet, it seems that this is not what many would desire for epistemology. The whole point, one can protest, is precisely to produce a normative framework that is over and above any sources of variation. This would be similar, thus, to the linguistic task of devising a universal grammar. Epistemic intuitions from all sources would provide the data for the epistemologist to abstract and organise. Comparative studies would be needed, appropriate taxonomies should be furnished, and so on and so forth. Ideally, the end product would be a theory of the competence that complemented with the pertinent performance factors would account for all the range of variation in epistemic judgements.

There is a problem with this view, though. There is a sense in which theories of particular grammars are not defeasible. Suppose we find in a remote part of England a community that shares the entire English vocabulary, but that uses postpositions instead of prepositions. It may not reasonable to conclude that "the woman red in is dancing you with" is an English sentence. Rather, we might conclude that "they do not speak English", i.e., that the peculiar community speaks a different language (call it Schmenglish) with its own particular grammar. Yet, a theory of universal grammars can be defeated by such findings. Suppose that it is a rule of our UG that interrogative sentences are not constructed by reversing the order of the words in a declarative sentence (cf. Pinker, 1994), and suppose we find a remote community that follows exactly this procedure for interrogative sentences. Then it would be reasonable to call for a revision of our UG. (Notice that taking a line similar to the English/Schmenglish case would amount to the absurd claim that the community in question "does not speak a (human) language"). It is precisely its universality aims that make the search for a UG prone to those kinds of revisions.

Thus, if the comparison between epistemology and linguistic is apposite, and we want to characterise the former as the search for a "universal epistemic grammar", then the conclusion is that epistemology is also revisable in the face of diverging epistemic intuitions. It is not enough to state that, say, Asians and Westerners have different "epistemic grammars"; we need, in addition, a general framework that encompasses the intuitions of those divergent grammars. The trouble is that while the lack of normative standards to prefer a language over another is no obstacle to the linguist, it seems that the epistemologist *must* decide between the divergent intuitions offered by different people. The (linguistic) UG has to offer a scheme that allows for the possibility of the distinct linguistic variations. The "universal epistemic grammar", in contrast, has to explain how the variations are possible, but also how and why some of them are better than others.

A second possibility about the nature of epistemic intuitions is that they are much like artistic or moral intuitions. But to argue this is not an easy task. First, variability is much greater in moral or artistic intuitions than in epistemic ones. There are not only marked differences between individuals from disparate cultural or socioeconomical groups, but even between individuals that belong to the same group. Second, while epistemic intuitions seem to be much more malleable by explicit formal training (e.g., taking epistemology classes), moral intuitions seem to be much more resilient to change from that kind of pressure. If epistemology lectures were just another form of indoctrination, then it is difficult to see why explicit formal indoctrination of that kind fails in the case of inculcating a particular set of moral values.

Let us consider now the third possibility that epistemic intuitions are like biological intuitions. On the one hand, like the latter, epistemic intuitions are not innate. Whatever concepts and beliefs we may attribute to newborn babies, it is unreasonable to impute epistemic judgements to them. We do not imagine babies going on in the business of trying to determine whether their mothers "really know" that they are wet. Epistemic intuitions are not explicitly acquired, at least for the vast majority of the mortals that do not attend epistemology courses. So, where do epistemic intuitions of epistemologically untrained individuals come from? An appealing hypothesis is that they come from *informal* enculturation, just like a substantial part of our ordinary bodies of knowledge. The common fellow does not attend epistemology seminars but he is engaged in social interactions in which people are judged to know something, behaviours are labelled as rational or irrational, beliefs are considered warranted or unwarranted, and so on.

Supposing, then, that epistemic intuitions are much like biological intuitions, it seems that epistemology is doomed. Because it is an outright feature of biological intuitions that they can be defeated: they do not have normative force. We can save the day, however, accepting that, even if biological intuitions do not have normative force, *biology does*. In other words, the reason why an intuitive judgement like, say, that fungi belong to the same broad class as plants, can be overruled, is that there is a systematised, reasoned and thoroughly elaborated corpus of biological findings. To be sure, any of the statements that belong to a theory of biology can be defeated as well, but they cannot be defeated without a good reason.

Namely, the fact that they clash with non-reflective folk intuitions is not such a good reason.

These considerations obviously appeal to the role of expertise in deferring our common biological intuitions. The point is now whether there is a similar role for expertise in epistemology. Weinberg et al. reject this possibility. They consider a number of objections which say that in their experiments they are probing the wrong sort of intuitions, because the right sort is accompanied by a sense of necessity, or require a minimum of reflection, or emerge after an extended period of discussion and reflection. In fact, all these objections can be regarded as variations in the single theme of the role of expertise. The idea is that some people may have a greater epistemological authority simply because they have devoted more time to think about epistemic affairs, an idea that they discard.

However, the surprising consequence of this view is that it implicitly endorses the assumption, shared by the "intuition driven romanticists", that intuitions are incorrigible in a relevant sense. Or to be more precise, in this view intuitions are malleable but incorrigible. They are malleable because they are subject to cultural and socioeconomical and other kinds of influences. But they are incorrigible because we are not entitled to say that a particular set of intuitions is wrong, on pain of committing the sin of ethnocentrism.

In our view, the possibility of scepticism tells against the idea of incorrigibility of our epistemic intuitions. If (at least some variety of) scepticism is correct, then claims about knowledge are mere loose talk. The most we would be able to reach is a state of belief. If scepticism were a strong position, then it would seriously undermine most of the epistemic intuitions which apparently are universal, namely, that knowledge constitutes a distinct class of mental state attributable to people. In fact, the extraordinary urge to fight scepticism may well be motivated by its threat to this strong and widely shared intuition. This does not prove the primacy of the intuition, only its strong appeal, which is undermined by the sheer possibility that scepticism might be right. But it is not possible to maintain at the same time both that epistemic intuitions are incorrigible, and that scepticism is a possibility, because one of the consequences of scepticism is that intuitions are not incorrigible after all. We are not sure if somebody would like to use this as an argument against scepticism. We regard it as an argument against the incorrigibility of epistemic intuitions, because the *possibility* of scepticism must be taken for granted in any sensible epistemological project.

Maybe the claim is not, after all, that epistemic intuitions are incorrigible. The claim may be that even if some of them were right and others were wrong, a problem remains to tell which are the good ones. This is something, the argument goes, for which we do not have an appropriate set of criteria. If intuitions diverge through, say, different social groups, what basis do we have to claim that an epistemological theory based on a particular set of intuitions has greater normative force than its alternatives?

But now the question looks much like the classical problem in philosophy of science of determining what makes a scientific theory a good theory. This is not to say that it is an easy problem, but it is not hopeless, in the sense that there are at least a number of alternatives to handle it. A possible answer to the question of when an epistemological theory is good, as in the biological case, lies in the success of the theory, in terms of accounting for a larger number of facts, or its internal coherence, or its predictive power, and so on. This is a theme to be pursued in philosophy of science terms, as a variation on the theme of how to evaluate different scientific theories. And this amounts to saying that to do epistemology is to do a particular kind of science. The best experts will be those that take into account the greater range of available data, including data from cognitive science, and that can explain the factors that make some epistemic intuitions right and others wrong. This would be good news for those that envision a naturalistic project for epistemology, whose aims and methods would be continuous with those from cognitive science.

5. FOLK EPISTEMOLOGY AND FOLK PSYCHOLOGY

We began part II talking about the relation between folk epistemology and folk psychology, and we want to conclude by drawing some morals that are common for both of them. First, we pointed out that folk epistemology shares its ontology with folk psychology. Yet, as we remarked in part I, the status of this ontology is very much in dispute. Folk epistemology inherits these problems. On the one hand, it may happen that typical epistemic notions do not carve epistemology at the proper joints. What we call knowledge may refer to a number of disparate states that might correspond to types of neural states, types of environmental states, or both. In addition, it may be the case that we are utterly wrong when we judge that somebody (or even ourselves) knows or believes or understands X. Phenomena like confabulation (Nisbett and Wilson, 1977) or self-deception (McLaughlin and Rorty, 1988) cast some doubt on the possibility that epistemic conditions for the attribution of knowledge can customarily be satisfied. If there are no clear and distinct conditions of success for the case of self-knowledge, and if attributions of knowledge depend to some extent on the own case, then the project of developing an epistemology based on folk epistemology seems to be in trouble.

Conversely, the reflections that we just made about the nature of epistemic intuitions can be made applicable to folk psychology. In other words, it might be the case that in both cases there is a core of "folk wisdom", a core that may well be evolutionarily fixed and universally shared; but it might also be the case that much of the discourse that is taken for folk psychology could be just like the discourse of epistemic intuitions, i.e., utterly subject to cultural, socioeconomical, psychosocial or other kinds of influences. If this is the case, the conclusion is not necessarily a relativistic position towards epistemology and psychology. Rather the conclusion is that both of them may be revisable in the face of empirical findings and theoretical developments obtained underneath a scientific umbrella. This umbrella is not other than cognitive science. In this case, much of the epistemological debate as well as much of the folk psychology debate will have the shape that debates take in philosophy of science, and they will have to be tackled in the terms and limits proposed by this discipline.[6]

REFERENCES

BARNES, B. & BLOOR, D. (1982) "Relativism, Rationality and the Sociology of Knowledge", en Hollis, M. & Lukes, S. (Comps.) (1982) *Rationality and Relativism*, Oxford: Blackwell.

BLOOR, D. (1981) "The Strengths of the Strong Programme", *The Philosophy of Social Sciences* **11**: 199-213.

BLOOR, D. (1991) "Ordinary Human Inference as Material for the Sociology of Science", *Social Studies of Science* **21:** 129-39.

BECHTEL, W. (1995) "What Should a Connectionist Philosophy of Science Look Like?", in R. N. McCauley (ed.) *The Churchlands and Their Critics*, Oxford: Blackwell.

BURGE, T. (1979) "Individualism and the Mental", *Midwest Studies in Philosophy*, **4**: 73-122.

CHURCHLAND, P. M. (1989) *A Neurocomputational Perspective: the Nature of Mind and the Structure of Science*, Cambridge, MA: MIT Press.

COSMIDES, L. & TOOBY, J. (1994) "Origins of Domain Specificity: The Evolution of Functional Organization", in L. A. Hirschfeld and S. A. Gelman (eds.) *Mapping the Mind*, Cambridge: Cambridge University Press, pp. 85-116.

[6] This paper is part of the research projects UPV/EHU 109.109-HB 078/99 of the University of the Basque Country and BFF200203842 of the Ministry of Science and Technology. Fernando Martínez-Manrique enjoys a post-doctoral scholarship BFI00.44 of the Departmento de Educación, Gobierno Vasco. We would like to thank the audience at ICCS'01 and an anonymous referee for the comments that helped to improve this paper.

CUMMINS, R. *et al.* (2001) "Systematicity and the Cognition of Structured Domains", *The Journal of Philosophy*, XCVIII, pp. 167-185.

DAVIES, M. & STONE, T. (1995) "Introduction", in M. Davies and T. Stone (eds.) *Folk Psychology*, Oxford: Blackwell, pp. 1-44.

DENNETT, D. C. (1987) *The Intentional Stance* Cambridge, MA: MIT Press.

DENNETT, D. C. (1991) "Two Contrasts: Folk Craft versus Folk Science and Belief versus Opinion", in J. Greenwood, ed. *The Future of Folk Psychology*, Cambridge: Cambridge University Press.

FODOR, J. (1981) "Three Cheers for Propositional Attitudes", in J. Fodor *Representations*, Cambridge: MIT Press.

FODOR, J. (1983) *The Modularity of Mind,* Cambridge, MA: MIT Press.

FODOR, J. (1987) *Psychosemantics. The Problem of Meaning in the Philosophy of Mind*, Cambridge, MA: MIT Press.

FODOR, J. (2000) *The Mind Doesn't Work That Way. The Scope and Limits of Computational Psychology,* Cambridge, Mass: MIT Press.

FODOR, J. & PYLYSHYN, Z. W. (1988) "Connectionism and Cognitive Architecture: A Critical Analysis", *Cognition,* **28**: 3-71.

FULLER, S. (1993) *Philosophy of Science and Its Discontents*, New York: The Guilford Press.

GETTIER, E. L. (1963) "Is Justified True Belief Knowledge?", *Analysis*, **23**: 121-123.

GIERE, R. N. (1988) *Explaining Science. A Cognitive Approach*, Chicago: The University of Chicago Press.

GIERE, R. N. (ed.) (1992) *Cognitive Models of Science*, Minnesota Studies in the Philosophy of Science, Vol. XV, Minneapolis: University of Minnesota Press.

GIERE, R. N. (1994) "The Cognitive Structure of Scientific Theories", *Philosophy of Science* **61**: 276-296.

GOLDMAN. A. (1986) *Epistemology and Cognition*, Cambridge: Harvard University Press.

GOPNIK, A. & MELTZOFF, A. (1997) *Words, Thoughts, and Theories*, Cambridge, MA: MIT Press.

HARMAN, G. (1986) *Change in View,* Cambridge, MA: MIT Press.

HORGAN, T. & TIENSON, J. (1994) "A Nonclassical Framework for Cognitive Science", *Synthese,* **101**: 305-345.

HORGAN, T. & TIENSON, J. (1995) "Connectionism and the Commitments of Folk Psychology", in *Philosophical Perspectives, 9,* pp. 127-152.

HORGAN, T. & TIENSON, J. (1996) *Connectionism and the Philosophy of Psychology*, Cambridge, MA: MIT Press.

JOHNSON-LAIRD, P. N. (1983) *Mental Models*, Cambridge: Cambridge University Press.

KUHN, T. S. (1970) *The Structure of Scientific Revolutions*, 2nd ed., Chicago: University of Chicago Press.

LAKATOS, I. (1970) "Falsification and the Methodology of Scientific Research Programmes", in Lakatos and Musgrave (eds.), 1970, *Criticism and the Growth of Knowledge*, Cambridge: Cambridge University Press, pp. 91-196.

LAKATOS, I. (1971) "History of Science and Its Rational Reconstructions", in Buck and Cohen (eds.) (1971) *PSA 1970,* East Lansing: Philosophy of Science Association, pp. 91-135.

LAUDAN, L. (1987) "Progress or Rationality? The Prospect for Normative Naturalism", *American Philosophical Quarterly* **24/1**: 19-33. Reprinted in L. Laudan (1996) *Beyond Positivism and Relativism: theory, method and evidence*, Boulder, CO: Westview Press.

MCLAUGHLIN, B. P. & RORTY, A. O. (eds.) (1988) *Perspectives on Self-Deception* Berkeley, CA: University of California Press.

MARR, D. (1982) *Vision,* San Francisco: Freeman.

NERSESSIAN, N. J. (1992) "How Do Scientists Think? Capturing the Dynamics of Conceptual Change in Science", in R. Giere (ed.) 1992, pp. 3-42.

NERSESSIAN, N. J. (1995) "Opening the Black Box: Cognitive Science and History of Science", *Osiris* **10**: 194-214.

NISBETT, R. & WILSON, T. (1977) "On Saying More than We Can Know", *Psychological Review,* **84**: 231-259.

PAPINEAU, D. (1993) *Philosophical Naturalism* Oxford: Blackwell.

PAPINEAU, D. (1996) "Précis of *Philosophical Naturalism*", *Philosophy and Phenomenological Research* **LVI**: 657-665.

PINKER, S. (1994) *The Language Instinct,* New York, NY: Harper.

POPPER, K. R. (1972) "Two Faces of Common Sense", en K. R. Popper *Objective Knowledge,* Oxford: Oxford University Press.

PUTNAM, H. (1975) "The Meaning of 'Meaning'", en H. Putnam (1975) *Mind, Language, and Reality. Philosophical Papers II,* Cambridge: Cambridge University Press.

RAMSEY, W., STICH, S. P. & GARON, J. (1991) "Connectionism, Eliminativism, and the Future of Folk Psychology", in W. Ramsey, S. P. Stich, and D. E. Rumelhart (eds.) *Philosophy and Connectionist Theory,* Hillsdale, NJ: Erlbaum, pp. 199-228.

ROUSE, J. (1998) "New Philosophies of Science in North America — Twenty Years Later", *The Journal for General Philosophy of Science* **29**: 71-122.

SAMUELS, R. (1998) "Evolutionary Psychology and the Massive Modularity Hypothesis", *British Journal for the Philosophy of Science,* **49**: 575-602.

SCHOLL, B. J. & LESLIE, A. M. (1999) "Modularity, Development and 'Theory of Mind'", *Mind and Language,* **14**: 131-153.

SELLARS, W. (1963), *Science, Perception and Reality.* London: Routledge and Kegan Paul.

SMOLENSKY, P., G. LEGENDRE & MIYATA, Y. (1992) "Principles for an Integrated Connectionist-Symbolic Theory of Higher Cognition", Computer Science Department, University of Colorado at Boulder, Report CU-CS-600-92, pp. 41-42.

SOLOMON, M. (1992) "Scientific Rationality and Human Reasoning", *Philosophy of Science* **59**: 439-455.

SOLOMON, M. (1994) "Social Empiricism", *Noûs* **28**: 325-342.

SOLOMON, M. (1996) "Commentary on Alison Gopnik's *The Scientist as a Child*", *Philosophy of Science* **63**: 547-51.

SUCHMAN, L. (1990) "Representing Practice in Cognitive Science", en Lynch, M. & Woolgar, S. (eds.) (1990) *Representation in Scientific Practice,* Cambridge, Mass.: MIT Press, pp. 301-322.

THAGARD, P. (1988) *Computational Philosophy of Science,* Cambridge, MA: MIT Press.

THAGARD, P. (1992) *Conceptual Revolutions.* Princeton: Princeton University Press.

TOOBY, J. & COSMIDES, L. (1992) "The Psychological Foundations of Culture", in J. H. Barkow, L. Cosmides, and J. Tooby (*eds.*), *The Adapted Mind: Evolutionary Psychology and the Generation of Culture,* Oxford: Oxford University Press, pp. 19-136.

WEINBERG, J. M., NICHOLS, S. & STICH, S. P. (forthcoming) "Normativity and Epistemic Intuitions", *Philosophical Topics.*

Chapter 8

MUSIC, LANGUAGE AND COGNITION: WHICH DOESN'T BELONG?

Peter Kivy
Rutgers University, New Brunswick

INTRODUCTION

When the topic of "Music, Language and Cognition" was first proposed to me by the organizers of this Colloquium, I at once was reminded of the game where a child is shown, for example, pictures of an apple, a banana and trumpet and asked "Which one doesn't belong?" I assume you all know the answer.

But which doesn't belong in the triad of music, language and cognition? If you answer that they *all* belong, as some people might, then you are in disagreement with me. For my answer is: *language* doesn't belong. Music is, indeed, language-like in certain respects. Nevertheless, it is not language: it is not a language or part of a language. And thinking it is any one of these things has caused a good deal of confusion.

Certainly, though, music and *cognition* belong together. For, on my view, anyway, the enjoyment and appreciation of music are deeply involved with the cognitive processes. As I put the point some years ago, "music is not a stimulus...: it is an object of perception and cognition, which understanding opens up for...appreciation."[1]

[1] Peter Kivy, *Music Alone: Philosophical Reflections on the Purely Musical Experience* (Ithaca: Cornell University Press, 1990), p. 41.

K. Korta and J. M. Larrazabal (eds.), Truth, Rationality, Cognition, and Music, 157-174.

My talk today will be solely concerned with absolute music, that is to say, pure instrumental music without text, title, or program; and it will naturally fall into two parts. In the first I want to explore the music-language analogy, emphasizing the while that it *is* an *analogy*: that music is not language, not a language, not part of a language. In the second part I want to outline *some* of what I mean when I say that the enjoyment and appreciation of music are concept-laden activities: *cognitive* activities, if you will. Whether such cognitive activities are what cognitive science is all about I do not know. But I dare say this is the place to find out. First, then, to the music-language connection.

MUSIC AND LANGUAGE

When Franz Joseph Haydn, "Papa" Haydn to this friends, decided, in 1790, at the advanced age of 58, to make an extended trip to England, Mozart is supposed to have exclaimed: "Oh, Papa, you have had no education for the wide world, and you speak so few languages". Haydn's legendary reply was: "But my language is understood all over the world."[2]

What is it about this famous statement that sounds so true? And what is it that is so palpably false?

In answering these questions there are three aspects to Haydn's claim that we should look at: what he would have meant by his music, which is, of course what he is talking about, constituting his language; closely related to that, what he would have meant by someone's understanding it; and, finally, why he would have felt so confident that his music was understood by listeners "all over the world," "the wide world", as Mozart called it—a fairly audacious claim, one feels compelled to say.

Well, to start with the last aspect of Haydn's remark, his reference to the whole world, it becomes far less grandiose when one considers what the "world" of Haydn and Mozart was. It consisted, after all, of Austria, Germany, Bohemia, France, Italy and England; certainly not much more than that: that was it. And it does not seem unreasonable to think that *that* "world" could "understand" Haydn's music (although we do not yet quite know what "understand" might mean in this context). In other words, an Englishman who understood no German could still "understand" the *musical* "language" that Haydn "spoke". It was, in other words, an "international language": the true *lingua franca*.

[2] Karl Geiringer, *Haydn: A Creative Life in Music* (London: George Allen and Unwin, 1947), p. 89.

Of course *we* all know perfectly well that had Haydn taken his music to Japan, or China, or India, he would have found in those places no understanding of his musical "language", nor would *he*, genius though he was, have understood the musical "languages" of Japan, China or India. And in that respect, then, music is language-like. For just as someone brought up in a French-speaking country understands French but not German, so someone brought up in China understands Chinese music but not the music of Haydn and Mozart.

Someone might object here that the language analogy in this respect is not a very good one, since the French understand French but no German, the Germans understand German but no French, and so on, whereas *all* Europeans understand Haydn, which is to say, Viennese classical music. Well, I am not sure just how important this difference is, or what the reasons for it are. One reason, obviously, for the broader understanding of European music, as opposed to European language, must be that natural languages have a semantics as well as a grammar. You have to know what French words mean to understand French. But you don't have to know what the themes (or whatever analogue to words you choose) of a classical symphony mean to understand *it*. They don't *mean* anything. One way of putting this is to say that, unlike natural languages, music of the kind Haydn wrote has a "syntax" without a semantics.

Since you are not reading this lecture, you cannot see that in referring to the "syntax" of music, I have put "syntax" in quotation marks: scare quotes as they are sometimes called. I do this to alert the unwary to the fact that, at least so I believe, musical "syntax" may not be syntax in the literal sense, but syntax-like, as music is language-like but not language.

It looks, then, as if music may have been understood in the whole of Haydn's world, while German and English were not, because to understand Haydn's music you only needed to know a syntax, whereas to understand German or English or French you needed to know a semantic as well, and a different one for each language. Furthermore, the grammars of these natural languages differ, even though there are aspects they have in common, so that if you understand French grammar, say, you will also understand at least some of the principles of English and German grammar as well. (I am leaving out here the whole issue of Chomskian deep structure, and its supposed analogue in tonal music, as laid out by Lerhdhal and Jackendoff).

In any case, we can reach the reasonable conclusion, to return to Haydn's response to Mozart, with which we began these reflections, that Haydn was right in claiming that his music was understood all over the world, but only because the world to which he referred was *his* world, a very restricted, homogeneous world, which shared his musical culture. (Multiculturalism was not something Haydn and his century had to worry about). And in that unlike French, or German, or English, Haydn's music *was* understood throughout his world, and still is, it is very *un*-language-like. On the other hand, in that it was not understood in China, Japan or India, where very different musics flourished, it *is* very language-like. For it is of the nature of natural languages to be understood only by those who, one way or another, have learned them. And the common way to learn a natural language is to be born into it and be immersed in it throughout one's life. It is the same, needless to say, with music, whether it be Chinese music, Japanese music, Indian music, or the music of the West. You aren't likely to be aware that you are learning it. You just *are*.

But what, after all, does it really mean to say that all the world, Haydn's world, *understood* Haydn's music? What does understanding such music amount to? If I am right—and I am not alone in thinking this—that music has no vocabulary, has no semantic content, then we clearly cannot mean that music is understood in the way French or German or English is. Understanding the meanings of words is, after all, a substantial part of linguistic competence. You have to have a vocabulary. So what could we plausibly construe Haydn as saying when he averred that the whole world *understood* his language, which is to say, his *music*?

In the second part of this lecture, when I come to talk about music and cognition, I will go into this matter a little more deeply. But for now it will suffice to say that what it means to understand music is to enjoy and to appreciate it for the right, relevant reasons: that is to say, to enjoy and appreciate in it those aspects of it that the composer intended you to enjoy and appreciate, and to enjoy and appreciate them in the way or ways intended. Thus, for example, if I enjoy and appreciate the way Bach combines the two themes together, in a fugue, after developing each of them separately, or the way Beethoven brings in the horn, "in the wrong place", just before the recapitulation, in the first movement of the *Eroica*, I am displaying enjoyment and appreciation of the music; and that is what I mean by understanding it. But if I am completely unacquainted with the music of Southern India, and am enjoying the exotic jangle of the instruments while eating a curry in an Indian restaurant, I am neither enjoying nor appreciating that music. And that amounts to saying that I am not understanding it. I am certainly enjoying and appreciating *something*; but not *it*.

So far, then, in our exploring of the music-language analogy, we have reached the following conclusions. Music is language-like in that it is *not*, as Haydn is sometimes quoted as having said, an international language. The music of India, like the language of Germany, is not generally understood beyond national boundaries. The understanding of Indian music, like the understanding of the German language, is learned, not innate: learned, to be sure, naturally and effortlessly, within a culture, at least to a certain degree of minimal competence; but hardly cross-culturally understood, the way pictorial representations apparently are. Of course a German can learn to understand Indian music, as an Indian can learn the German language. And that too marks music out as language-like.

Furthermore, music is language-like in having something *like* a syntax. On the other hand, it is surely far from language-like in its total lack of a semantic component. And that lack alone is sufficient for concluding that music is not language, a language, part of a language, or however you want to put it.

Nevertheless, there is *another* aspect of Western music, possessed by it since the beginning of the seventeenth century, that has been so frequently adduced as evidence of its linguistic credentials that no discussion of the music-language analogy would be complete without considering it. The two most frequent characterizations of music as language are that music is an "international language" and that music is "the language of the emotions". We have looked at the former claim already, and I turn now to the latter one. They are not, it turns out, unrelated, as we shall see.

The notion that music is an emotional (as opposed to a conceptual) *language* became so wide-spread by the end of the eighteenth century that it amounted to more or less a philosophical consensus. Kant, to adduce a prominent example, simply took it for granted. Music, he asserted without any argument at all, is a speech or language of the emotions: *Sprache der Affekte*.[3] Where did this view originate, and what, if any, plausibility attaches to it?

To answer the former question I must mention two momentous events in music history that transpired in the second half of the sixteenth century. The first, not merely an event of musical significance, but of world significance as well, was the so-called counter-reformation, the second the invention of the musical art form we know as opera.

The counter-reformation was the attempt on the part of the Roman Catholic Church to, in effect, clean house, in response to the criticism of the Protestant reformers. What this had to do with music was that the music of the liturgy, by the beginning of the sixteenth century, the heyday of Flemish polyphony, had become so elaborate and complicated that the text was all but unintelligible. And it was the *text*, the church fathers reminded the composers, that was the whole point of the exercise.

The villain was the polyphony itself: the combining of five or more long, sinuous melodies in such way as to prolong a sentence or phrase of text beyond the point where a listener could tell what words were being sung.

[3] Immanuel Kant, *Kritik der Urteilskraft* (Hamburg: Felix Meiner Verlag, 1959), p. 186 (53).

And if you couldn't understand the words, obviously the religious sentiments they expressed would be lost to you as well, and the very purpose of your presence at worship was defeated.

The fate of Catholic liturgical music, along with the many other issues of the counter-reformation, were debated at the Council of Trent, which sat from 1545 to 1563. Here, in part, is the admonition that was directed by the Council to the composers. "The whole plan of singing in musical modes should be constructed not to give empty pleasure to the ear, but in such a way that the words may be clearly understood by all, and thus the hearts of the listeners be drawn to the desire of heavenly harmonies, and the contemplation of the joys of the blessed."[4]

How was this admonition to be executed in practice? Well, if the best way of making words intelligible to a listener is to speak them, then the best way of making them intelligible to a listener, when they are sung, is, one might well suppose, to *represent* the speaking voice in music. What the Council of Trent accomplished, perhaps unwittingly, was to make of vocal music a representational art. I cannot prove that this is what the so-called Post-Tridentine composers of Catholic liturgical music had on their minds. But when I listen to their works, especially those of Palestrina, that is what the music sounds like to me: the representation in music of human speech.

In any case, if the representation of the human speaking voice was implicit in the practice of composers of liturgical music in Italy, after the Council of Trent, it quickly came to be *explicit*, with a vengeance, in the writings of the so-called *Camerata*, in Florence, at the close of the sixteenth century: the group of poets, theoreticians, composers and noblemen who were in the process of inventing musical theatre: *drama per musica,* as it was originally called, opera, as we call it now. Their idea was that the only way you could emotionally move the listener—and the moving of the emotions was the most important aim of music, so they thought—was to represent in your music the declamation of a speaker whose utterances are being made under the influence of some powerful emotion. As Giulio Caccini, one of the earliest composers of opera, described his practice, "for, unless the words were understood, they could not move the understanding, I have endeavoured in those [,] my late compositions to bring in a kind of music by which men might, as it were, talk in harmony…"[5]

The basic idea behind the musical representation of impassioned speech in early opera was that the listener would empathize with the character thus expressing herself and thereby experience the very emotion being expressed.

[4] Quoted in Gustave Reese, *Music in the Renaissance* (New York: Norton, 1954), p. 449.

[5] Oliver Strunk (ed.), *Source Readings in Music History: From Antiquity through the Romantic Era* (New York: Norton: 1950), p. 378.

I doubt that the empathy theory is in very good repute these days, and for our purposes it is irrelevant. What is important to observe is that in lock step with the establishment in Western music of the major-minor tonal system, in the seventeenth century, there evolved a whole arsenal of musical themes and harmonic techniques whose emotive character became instantly recognizable to the competent listener. Furthermore, this fundamental recognizability endures to this day. As long as the composer remains within the major-minor tonal system, he or she can still rely on the competent listener to recognize when the music is sad, when joyful, when ominous and so on. These emotive building blocks are the stock in trade of the film composer, the composer of popular songs and are part and parcel of the whole classical music repertoire from the beginning of the seventeenth century to the breakdown of tonality in the twentieth.

I have avoided doing so on purpose, but it is almost impossible to refrain from calling these emotive building blocks I have been alluding to an emotive "vocabulary," making up an emotive musical "language". And as a *façon de parler* it is perfectly harmless. Music is certainly language-like in having these universally recognizable emotive qualities. Where we get into trouble is taking the language analogy too seriously, which is to say, too literally. It is one thing to say that this passage of music is sad, that one triumphant. It is a different thing, and a profound mistake, to say that the one passage *means* "sadness," the other "triumph," and that the music is *saying* something about sadness and triumph. *That* is not a part of what music does.

There has been lively debate among philosophers over the last fifty or so years about what exactly we mean when we describe music in emotive terms. But a general consensus has emerged from this debate that we are *not* ascribing dispositions to the music to arouse emotions in us but, rather, that we are ascribing phenomenal, heard properties *to* the music. As the American philosophers O.K. Bouwsma once wittily put it, "the sadness is to the music rather like the redness to the apple, than it is like the burp to the cider."[6]

This being said, we might now look, finally, at the relation I suggested there might be between the notion that music is an international language and the notion that it is a language of the emotions. Might not the latter feed the former? For if emotions are universal to the species, in expression and experience, common to all cultures, then if music *were* a language of the emotions, it would, it might be argued, be a cross-cultural, which is to say, an international language.

[6] O.K. Bouwsma, "The Expression Theory of Art," in *Philosophical Analysis*, ed. Max Black (Ithaca: Cornell Universtity Press, 1950), p. 100.

There was a time when this argument would have been rejected on the grounds that emotions and their expression are *not* universal to the species but culturally conditioned: culture specific. However there is mounting empirical evidence that at least the basic emotions and their expression are not matters of culture and conditioning; rather, biological endowments.

What *is* wrong with the argument is its reliance on the false assumption that the emotions *in the music* are universally recognized across cultural boundaries. They are not, because they are aesthetic qualities of the music. Being so, they can only be perceived by musically competent listeners, and musical competence, as we have already seen, is not species-universal but culture-bound. An Indian, unacquainted with Western musical culture and idioms, cannot tell us that a passage in one of Haydn's symphonies is sad rather than happy, not because she experiences sadness and happiness differently from the way we do, or expresses them differently. She cannot perceive that the music is sad or happy because she cannot yet perceive *the music*, of which the sadness or happiness are aesthetic components. It is Western musical sadness and happiness, not Western sadness and happiness that are opaque to her.

The notion that music is a language of the emotions, then, like the notion that music is an international language, has a kernel of truth in it: it reveals to us one of the ways music is language-like. It *is* language-like in that the competent listener to Western music—and I make no claims in this regard about any other music—can recognize the emotive qualities of the music in a consistent manner. And in this regard it is language-like too in that the ability to "read" the musical emotions, like the ability to read French or German, is not innate or cross-cultural. Just as you must learn to read French or German, you must learn to "read" the emotion in Western music. Music is not a language or the language of the emotions. But its emotive character makes it language-like in that respect.

I conclude this section of my talk, then, with the general observation that music is *not* a language but is *language-like* in the ways I have just enumerated. In the triad, music, language, cognition, language is the one that doesn't belong. But music and cognition, on the other hand, are an inseparable pair. And I go on, now, in the second section, to make some observations about their cordial relations.

MUSIC AND COGNITION

There has been prominent in our thinking about music, absolute music, since the eighteenth century, what appears to me to be a false dichotomy between the view that music is content-less, and therefore concept-less, and the view that music must be concept-laden and therefore content-laden. Concept-less or content-laden have been the two alternatives on offer.

For contemporary philosophers the *locus classicus* of the content-less therefore concept-less alternative is Kant's definition of music as "the art of the beautiful play of sensations."[7] And perhaps the same role is played by Schopenhauer for the content-concept view in his claim that music is an image or likeness of the will: *Abbild des Willens*.[8]

In the contemporary world, the ascribing of content to absolute music has, after long eclipse in the shadow of formalism, become a flourishing business. To give you an idea of what kind of thing I am referring to, I quote from a well-known practitioner.

> By following the events of the first movement carefully, the listener becomes engaged in a process yielding a truth at the end. The forces used here are genuinely dramatic ones. In strictly musical terms, the opposition can be reduced to a conflict between stability and instability, a process not unlike that of any significant dramatic work. But instead of using characters or ideas or beliefs, the symphonist embodies his conflict in musical gestures which, in an archetypal way, parallel human conflicts. In the conclusion of the first movement of [Symphony] No. 83, Haydn can be seen to be demonstrating a very fundamental but difficult truth: opposition is inevitable, and the highest form of unity is not the one which eliminates conflict. On the contrary, it is one in which opposing forces can co-exist. The best minds of Haydn's age aspired to tolerance, not dogmatism. It is precisely this message that can be heard in Haydn's late symphonies.[9]

Now if music possessed this kind of semantic or representational content, there would be at least one obvious sense in which the enjoyment and appreciation of music would be a cognitive process. For I think all will agree that the perceiving of semantic or representational content in a work of art is a cognitive process if ever there was one. And if it is true, as the author just quoted suggests, that "messages" and philosophical theses can be heard in symphonies, and other such forms of music, then clearly we are obliged to acknowledge that the appreciation of such music is a palpably cognitive affair.

Speaking for myself, and for at least a substantial number of others, such semantic and representational interpretations will not withstand critical

[7] Kant, op. cit., p. 180 (51).

[8] Arthur Schopenhauer, *Die Welt als Wille und Vorstellung, Sämtliche Werke*, Vol. I (Stuttgart: Suhrkamp, 1986), p. 359 (Book III, 52).

[9] David P. Schoeder, *Haydn and the Enlightenment: The Late Symphonies and Their Audience* (Oxford: Clarendon Press, 1990), p. 88.

scrutiny. But I have neither the time nor the inclination right now to present arguments for my skepticism, although I have presented them in other places. Rather, what I want to do in the time remaining to me is to confront the disjunction, no content, therefore no cognition, *or* cognition, therefore content. It is a false disjunction. There *is* a third thing. There *is* cognition without content.

There are, as I see it, *at least* three kinds of cognitive processes going on when someone like me, who perceives no semantic or representational content in music, enjoys and appreciates it. I shall call them, not very surprisingly, conscious cognition, self-conscious cognition, and non-conscious cognition. It is the first two that I want to talk about here; but I mention the third because I do not want to be taken as denying that there may well be processing going on, when we listen to and enjoy music, that we are not aware of, and that can fairly be described as cognitive.

Indeed, the idea that there is non-conscious cognition taking place in the enjoyment and appreciation of music is not a new idea at all but a very old one. For if I read him aright, it was what Leibniz was saying, when famously he wrote of music, "even the pleasures of sense are reducible to intellectual pleasures known confusedly. Music charms us, although its beauty consists only in the agreement of numbers, and in the counting, which we do not perceive but which the soul nevertheless continues to carry out, of the beats or vibrations of sounding bodies which coincide at certain intervals."[10]

As I say, it is not non-conscious cognition, however, that I want to talk about. And that may come as a disappointment to many here. But, quite frankly, I do not know *how* to talk about non-conscious musical cognition, non-conscious computation, as Leibniz apparently saw it. So I had much better stick to what I think I do know how to talk about, namely, conscious and self-conscious cognition.

I think I mean something pretty obvious by the distinction between conscious and self-conscious cognition. When I see a cat, I take it I am in the conscious state of seeing the cat. But I also take it that I am not necessarily conscious *that* I am seeing the cat.

However, I may also become aware of my seeing the cat. In other words, my conscious state may have as its intentional object the cat, or it may have as its intentional object my conscious state of seeing the cat. Assuming, as I do, an epistemic view of perception, I call my perceiving of the cat conscious cognition and my awareness of perceiving the cat, with my perceiving of the cat as the intentional object, self-conscious cognition. Both

[10] Gottfried Willhelm Leibniz, "Principles of Nature and Grace." In *Philosophical Papers and Letters*, trans. and ed. by Leroy E. Loemker (Chicago: University of Chicago Press, 1956), Vol. II, p. 1042.

kinds of cognition are at work in enjoying and appreciating music. Here is how I see them operating.

I begin with the commonsensical notion that the performance of a musical work, that is to say, the "sounding" of a musical work, is a series of "sound events". What the events are, and how many there are is indeterminate, depending as it does upon what descriptions they are perceived or understood under and how one individuates the sounding, event-wise. So to ask how many events a work-sounding is composed of is very much like asking how many "things" there are in a room. It all depends on how you carve it up.

But the basic point is this. Serious enjoyment and appreciation of classical music is an activity of perceiving, sometimes consciously, sometimes self-consciously, musical events under musical descriptions. Classical music is not and is not meant to be merely a pleasing stimulation of the auditory faculty. Furthermore classical music, when correctly listened to, does not gain its effect solely by non-conscious processing, as Leibniz seems to have believed, although such non-conscious cognition no doubt occurs, and is vital to the experience. The enjoyment and appreciation of classical music is concept-laden *and* content-less, then, in that the entertaining of concepts, both consciously and self- consciously, on the part of the qualified, attentive listener, is quite essential to its full enjoyment and appreciation, but the concepts entertained are purely musical concepts, even though they may be described in ordinary, non-musical language. In other words, the concepts are not expressive of semantic or representational content. The enjoyment and appreciation of classical music, then, is both a conscious and self-conscious cognitive process without being one in which the cognition is either of semantic or representational properties.

Now the problem many people, particularly the musically untrained, have with this position, so stated, is that, they believe, it over-intellectualizes the musical experience. Yes, it may be granted, the musically learned—musicians, music theorists, musicologist—listen with a whole array of music-theoretic concepts and terminology at the ready. No doubt they very consciously and self-consciously apply these concepts to music as they listen to it. But, so the argument goes on, the untrained classical music lover *just listens*. How can she listen cognitively, either consciously or self-consciously? She doesn't have the concepts that such cognition requires. Only the "expert" does.

There are two points that should be addressed here. The first is the claim that only the expert's listening is cognitive, concept-laden listening because only the expert has the concepts. The second is the distinction between conscious and self-conscious listening and how I think it gets played out in musical listening, both that of the experts and that of the laity.

I think the best way to address both points is by way of a musical example.

W.A. Mozart, String Quartet in A, K.464.

Now there are may notable events taking place in this short but musically dense composition (which some of you may be familiar with). And the first thing to say about it is that one has not heard this composition unless one has consciously heard at least some major portion of these events as the events they are. Someone, for example, who has just been vaguely aware of the music, as background to something else he is doing, or as part of the ubiquitous musical "white noise" that pervades airports and bus and railroad stations, these days, to relax the nail-biting traveler, under the assumption that classical music is boring, and therefore anxiety-reducing, has not "heard" *the music* in a perfectly well understood use of perception words. For I am using "hear" in "hear the music" in a sense that I referred to awhile back as the "epistemic" sense. In that sense when one hears something one hears it *as* something; one takes it to be something. In other words one hears *it* under some description. It is the intentional object of one's conscious state.

Now what *does* one hear in this miniature compositional *tour de force?* Well, let me just say what *some* of the things are that *I* heard in the first few measures. I heard the opening theme as two separate four-measure phrases, the second kind of "answering" the first. In measures nine to twelve I heard a variation of the theme's first phrase. But I also heard that the second phrase of the theme was being used to accompany the first phrase. In other words, the main theme is so constructed that the first phrase and the second phrase can be played together, in counterpoint. In measures thirteen to sixteen I heard the first four notes of the theme in stretto. That is to say, one instrument begins the theme, and before it has finished playing it, another instrument begins to play it: violin 1, then viola, then violin 2, then cello. At measure seventeen things get even more complicated. Now the second part of the theme occurs in a stretto, but violins 1 and 2 begin the theme together, in octaves, and after a measure, before they are finished with it, the viola and cello enter together, the viola playing the second phrase of the theme while the cello plays the second phrase of the theme in inversion: that is, where the original theme ascends, it descends, and vice versa. So whereas the first phrase of the theme can be played with the second phrase as its accompaniment, we now discover that the second phrase of the theme can be combined, in stretto, simultaneously, with itself and its inversion.

These are some of the events I heard in the first twenty measures of the movement. There are many more events of this kind in the rest of it. But it would be pointless to go on enumerating them. I am sure you have gotten the general idea. But I do want to mention one further thing that I heard before I go on. So far I have been talking about what might be called the "internal" happenings in Mozart's minuet. However, the piece also follows a well known plan or pattern, if you will. It is constructed of two sections. The first section modulates to the dominant key of E. The second section, twice as long as the first, eventually modulates back to the tonic key of A. Each section is repeated. That is the general, very familiar plan or pattern working itself out as I listened, the while, to the internal events, some of which I have just described. And, of course, hearing that plan or pattern helped me to orient myself to the internal events I was hearing within it. I had the map.

Now when I say that I enjoy and appreciate this minuet of Mozart's, I mean, in part, that I enjoy perceiving these musical events playing out the general plan of the work. I enjoyed discovering them in my first hearings of it, and I enjoy rehearing and re-discovering them now, as well as, of course, sometimes discovering new events that I had not heard before. So what I mean by the enjoyment and appreciation of music of this kind being conscious cognitive processes is, in part, illustrated by the way I listen to and enjoy the Mozart minuet. For, I take it, it is obviously a conscious cognitive activity to hear that the theme of the minuet is in two phrases, that the first phrase is treated in stretto, beginning at measure thirteen, that the second phrase is combined with its inversion at measure eighteen, and so on. And to hear that these things are happening in the way I described I of course must possess the concepts of theme, phrase, stretto, inversion, counterpoint, and so forth. And so that is what I mean when I say that listening to music is concept-laden: that, furthermore, enjoying and appreciating music is concept-laden, but the music content-less. For the concepts that I deploy in enjoying and appreciating music are not its content: are not concepts that the music expresses. The music is not saying "stretto", "inversion," and so on.

But it is just here that the charge of *over-intellectualizing* the musical experience is liable to be made. For, so the argument goes, lots of people, genuine, serious music lovers all, enjoy and appreciate this minuet without possessing any of the concepts I have mentioned. They don't know what a stretto or an inversion is. They don't have any musical training. So how can they have any musical concepts at all to apply to the music? The expert's or trained amateur's listenings to music may be concept-laden, but the average music lover's are concept-*less*. Does that make the average music lover completely inconsequential from the aesthetician's point of view? That would be a hard saying indeed.

There is an element of truth in the charge. It is quite correct to say that the average, untrained music lover does not possess the music-theoretic concepts of the expert or trained amateur. Where it goes wrong is in suggesting that *therefore* the average, untrained music lover does not possess any musical concepts at all, and, as a consequence, is a completely "mindless" listener. Musical events need not be described in music-theoretic terms or be understood only under music-theoretic concepts. For many purposes ordinary language and non-technical concepts can suffice for rich enjoyment and appreciation.

So what does the serious but untrained music lover hear in Mozart's minuet? In the non-epistemic sense she hears, of course, the same musical events that the trained listener hears. But in the epistemic sense she hears some of these events under different descriptions from those of the expert, perhaps less satisfactory descriptions, and misses hearing some of them altogether.

Thus, for example, where I hear a stretto, the untrained listener will not hear what is going on under that description but may very well hear that *something* unusual is going on which she may want to describe as "crowding" or "clumping" or something like that. But she may well miss the inversion of the theme altogether, under any description, because I think it is hard to hear the inversion of a melody as a version of that melody at all unless the technique is pointed out and conceptualized. Anyway, the general idea is that music is describable in terms other than those of a music-theoretic kind, and can be heard under concepts other than those of a music-theoretic kind; and that, it appears to me, is sufficient to answer the charge of over-intellectualizing the musical listening experience, if it is a charge motivated by my use in my original description of that experience of music-theoretic terms and concepts.

But suppose the skeptic continues to press his claim, albeit in a somewhat scaled down version. He will say something like this. Granted there are *both* trained *and* lay listeners who enjoy and appreciate music in the cerebral, concept-laden, cognitive way you have described, nevertheless there *also* are large numbers of listeners who enjoy and appreciate music in a purely "mindless" way—at least mindless in the sense that they are not consciously or self-consciously perceiving musical events: rather they are simply pleased by the sounds, bathing, so to speak, in their fragrance, without a thought in the world *about* the music, although, of course, their minds (or brains) are very busy in cognitive and conceptual processes, *à la* Leibniz, of which these "mindless" listeners are completely unaware.

Now with this skeptical response to my views I have no quarrel at all. Doubtless it is true that there *are* mindless listeners, so-described by the skeptic. Nor am I here to read them a sermon on how they ought to change their mindless ways and listen to music the way I and my friends do. There is, indeed, music designed for mindless listening, and it pervades our aural space. The music of which I speak, however, the music of Bach and Haydn and Mozart, of Beethoven and Mendelssohn and Brahms, of Mahler and Schoenberg and Strawinsky, was not meant to be mindlessly enjoyed and appreciated, but enjoyed and appreciated in something like the cognitive, concept-laden way I have been describing. Of course it can also have a payoff for mindless listening. That is beside the point, which is that when it is listened to as intended, it has a different payoff, and one that has been found rich and rewarding by three centuries of listeners, both trained and untrained in its mysteries. That is the kind of music, and the kind of enjoyment and appreciation I am trying to understand. That there is other music, other kinds of enjoyment and appreciation, I am not here do deny, although if Clive Bell is to be credited, those so-called "mindless" listeners of which I have been speaking, at least some of them, anyway, may very well at least be aware that they are missing something important. As Bell put it, "About music most people are as willing to be humble as I am. If they cannot grasp form and win from it a pure aesthetic emotion, they confess that they understand music imperfectly or not at all."[11]

[11] Clive Bell, *Art* (New York: Capricorn Books, 1958), p. 31.

This brings us to the second point I said it was necessary to address: the distinction I have been assuming all along between conscious listening and self-conscious listening. I turn to that now, in conclusion.

It seems to me to make perfect sense to say that someone is aware of something in his visual or aural field but is not aware of his being aware. In the musical case, to return to my previous example, I may be perceiving that the second phrase of the theme is now accompanying the first phrase without being consciously aware that I am so perceiving. However, it sometimes happens in my own listening, and I am sure in other people's, that I start thinking about what I am perceiving as I am perceiving it; in this case I start thinking about the fact that the second phrase is now accompanying the first, and perhaps marvel over the skill and workmanship displayed. This is what I call self-conscious cognitive enjoyment and appreciation.

I believe that this kind of self-conscious, cognitive activity is very common among serious, sophisticated listeners. And I think it is common not merely among musically trained listeners, because I take it that a musical listener can be sophisticated to a high degree *without* being musically trained.

Now it may be—it very probably is—the case that technical musical training increases the tendency to listen to music self-consciously: to think about what one is hearing while one is hearing it. I know it increased that tendency in me. Is that a good result? Is it better to hear music self-consciously as well as consciously? I think it has been in my experience. Certainly it seems to add a layer of pleasurable experience to the enterprise.

It may be objected to what I am suggesting, however, that thinking about what one is hearing while in the process of listening to the music, even though it does occur, is not only not essential to the musical experience but downright destructive of it. If you think too much about what you are hearing, it will, so the objection goes, distract you from the hearing itself.

My own view is that you can't think too much about what you are listening to when listening to music. What I do think is that thinking too much frequently is confused with thinking about the wrong things. A composer may become so intrigued with a musical event she is perceiving that she will start thinking about how to use it in her own work. An oboist or violinist may become so intrigued with the techniques of players he is listening to that he starts thinking about his own playing technique. These are not cases of thinking too much about the music but cases of ceasing to think about the music altogether. It is, I suppose, thinking too much. What it is not is thinking too much about *the music*. Even though they are musical thoughts, they are as irrelevant to and distracting from musical listening as letting oneself think about the poetic or philosophical ideas that the music might suggest. Again in Clive Bell's words, "Tired or perplexed, I let slip my sense of form...and I begin weaving into the harmonies, that I cannot grasp, the ideas of life."[12]

I have been arguing, then, that there are two kinds of conscious cognitive activity going on in serious, concentrated listening to classical music. There is the conscious hearing of musical events, hearing *that* certain musical things are going on. And there is what I have been calling self-conscious hearing: hearing *that* certain musical things are going on, and, the while, thinking about these goings on and about your hearing them. These together seem to me to play a very large role in serious music listening. This is not to deny that there undoubtedly also is cognitive processing going on of which I am completely unaware while I am listening to music. I have said nothing of substance about such non-conscious cognition because, as I said before, quite frankly, I don't know how to do that. I have simply spoken about that of which I *can* speak. What I can't speak of, I can't speak of; and there's an end on that. But there probably *are* folks at this congress who *do* know about what I have been calling non-conscious musical cognition. And from them I am anxious to learn what I can.

CONCLUSION

Let me press on now to my brief and overdue conclusion. The topic proposed to me, and to Laird Addis, was Music, Language, and Cognition. I have argued that these are strange bedfellows, in that language should not be in the same bed. The enjoyment and appreciation of music is thoroughly imbued with conscious and self-conscious cognitive processes, and, no doubt, non-conscious processes as well. But the connection with music and language is purely analogical. Music is language-like in being culturally conditioned and being vaguely, but not literally describable as a "language of the emotions". That, so far as I can see, is where the connection ends. And a tenuous connection it is.

I have not really said here anything beyond what I have talked about and written about on various other occasions. But I am not apologetic for that. I am reminded of Groucho Marx's remark that there is no such thing as a new joke: a new joke is just an old joke someone hasn't heard yet. I told you all an old joke. My hope is that you haven't heard it yet, and can give me some new responses. Anyway, I presume that whenever one tries to re-express his views, as I have done for this occasion, the views change, if ever so slightly,

[12] Ibid., p. 30.

in the re-expressing. For, as Heraclitus said, you can't step in the same joke twice.

Chapter 9

MUSIC AND KNOWLEDGE

Laird Addis
The University of Iowa

There are, needless to say, multiple ways in which music *might* be related to knowledge; and no doubt, too, the actual ways are but a very small subset of the still probably finite number of imaginable ways. In the last few decades, cognitive science has often been invoked in discussion of putative musical knowledge—for example, by the philosopher Diana Raffman in her *Language, Music, and Mind* and, probably most notably, by the composer Fred Lehrdahl and the linguist Ray Jackendoff in their *A Generative Theory of Tonal Music*. One cannot doubt, even at this comparatively early date, the contribution that scientific psychology, done from a broadly "cognitivist" frame of reference, can make and has made to the understanding of the human mind in relation to music. If empirical science cannot tell us, as it surely cannot, what music is *worth* listening to, there remains the strong possibility that music psychology, including neurology and evolutionary biology, can tell us ever more why humans are so drawn to music and what they perceive *in* it and *from* it, and thus *why*, at least in some measure, they do evaluate music in the ways they do.

With respect to knowledge, cognitive science has had a special, though not exclusive, concern with sorts of knowledge that are called innate or tacit or unconscious as contrasted with those that are learned or introspectible or conscious. In that spirit, I propose in this paper to raise and give tentative answers to two questions in which knowledge that is innate or unconscious and having to do with music would seem to be at least strong possibilities. The first concerns knowledge that we *bring to* music, the other knowledge

K. Korta and J. M. Larrazabal (eds.), Truth, Rationality, Cognition, and Music, 175-190.

that we *get from* music–what I shall call, briefly, knowledge-*for* and knowledge-*from* listening to music. And it is well to keep in mind that, in another dimension of analysis, both knowledge-for and knowledge-from admit of a three-way distinction of knowledge-*that*, knowledge-*how*, and knowledge-*of,* as we shall see shortly.

It is well to begin these investigations with some reflections on the power and importance of music in human affairs. We have good reason to believe, on the "historical" side, that music may even have antedated that burst of human creativity—as evidenced for us mainly in tools and cave paintings—that occurred in our ancestors of about 40,000 years ago. Some even believe that our Neanderthal cousins had music in some form, perhaps learned from our own ancestors but perhaps—in an importantly different sense of 'from'—from innate impulses common to our genus even as far back as the half million years or so of the last common ancestor of them and us. It must be admitted, however, that what we know about the evolution of our species has so far not had much to tell us, in specifics, how there came to be organisms—ourselves—for whom music has the impressive value and power that it does have.

There is no reason, everything considered, to doubt that there does exist among the full facts of the matter some interesting ones about how the "music gene" came to be a favorable adaptation in *homo sapiens* or one of its ancestor species, although it must also be admitted that, *prima facia*, it would seem that the sorts of behaviors that constitute the practice of music would, if anything, be detrimental to purpose of favorable differential reproduction. But we shall see, although I shall not mention it again explicitly, that, looked at in a certain way and in the light of certain theories, the facts about the importance of music might be seen to fit more comfortably with the demands of evolutionary theory. Nor, by the way, do I mean to be begging any questions by speaking of a "music gene"; it is only a convenient way of referring to what it is in our biological nature, whatever it is, that—on a weak view—makes it *possible* for us to engage in musical behavior or—on a stronger view—almost *compels* that kind of behavior. But, as an initial point in favor of the stronger thesis: if a behavior can be said, usefully, to be *natural* to a species when, in one sense, it is highly likely to occur; or when, in another sense, its denial or inhibition correlates with increased mental anxiety and disorder, then we may say that music is surely natural to human beings in both senses.

But just what are some of the facts about the power of music for human beings? First of all, there are the anthropological facts including, above all, that music exists in every human culture known to historiography and social science. To be sure, some cultures, usually for religious reasons, greatly restrict music making—partly just because of its power—to narrowly circumscribed occasions, but no culture altogether lacks times and places for making music. This fact connects with a second fact of a sort we may also call anthropological having to do with the amounts of human and other resources that are devoted to the study, performance, appreciation, advocacy, composition, and teaching of music. The figures are not easy to come by, but we may reasonably believe that music consumes more time, effort, and money than all of the other arts—even very broadly construed—put together, and probably more than athletics, even in our sports-obsessed societies. Be that as it may, there is no doubt about the pervasive appearance and immense power of music in the lives of all peoples of all societies.

Yet a third anthropological fact concerns the wide variety of occasions on which music is used—presumably to enhance some feature of the occasion that is taken to be essential to it: the solemnity of a funeral, the merriment of a wedding feast, the excitement of a soccer match, the importance of a state ceremony, and so on and on. In fact, we humans use music in so many of our ceremonial and ritual occasions that in some societies the use of music virtually defines them as being just those sorts of occasions.

But, of course, underlying all of these anthropological facts—facts about the publicly observable role of music in mostly collective circumstances—are the *psychological* facts; that is, the connection of music with the individual human being. All such facts, whatever they are, presuppose that the subject is a conscious being, but one must not presuppose that all or even most of the important facts, or indeed any of the most important facts, have very much to do with what a person can or would report about his or her experience of music, that is, what goes on in the conscious, introspectible mind. This, however, is already somewhat to anticipate conclusions yet to be argued for. Yet this presupposition is attested to, if not always knowingly by their authors, by the now hundreds of studies that have been undertaken in the last century and a half under the general heading of *music psychology*. For music psychology, both experimental and theoretical, is the attempt to understand the nature of the human mind with respect to music—how we hear it, how we learn it, how we respond to it, and so on. And while it has not, at least explicitly and for the most part, seen its goal also to explain the overarching anthropological facts of the sort I just cited, it is obvious, insofar as we are "individualist" with respect to the nature and explanation of social phenomena, that it is in the science of music psychology, backed by neurology and evolutionary biology, that we shall locate ultimately most of the *empirical* facts that are relevant to explaining the power and importance of music for human beings.

But, as I hope to demonstrate in practice in the remarks that follow, these empirical facts, whatever they turn out to be, can be understood and assigned their comparative importance only in light of certain philosophical theories about the nature of, most importantly, consciousness, unconscious awareness, and knowledge—notions that practicing scientists as well as musicians tend to look on with some mixture of anxiety and skepticism. The first part of my claim is perhaps neither surprising nor controversial among philosophers who are interested in the idea, presuppositions, and claims of this or that aspect of cognitive science, but my experience, as a philosopher who has attended several conferences on music psychology, tends to confirm, alas, also the second part of my claim, which is itself a constituent of the larger, and often justified, skepticism about philosophy among practicing empirical scientists.

The other side of this complicated coin is the fact that many philosophers are suspicious of the idea of cognitive science in general or of the specific claims of this or that piece of cognitive science in particular. Some of this skepticism derives, no doubt, from political motives—and this despite Noam Chomsky's left-wing innatism—to the effect that the dignity and equality of human beings is somehow threatened by attempts to show that significant human behaviors are strongly grounded in innate mental structures or dispositions. For this is taken to suggest either that these behaviors are not, to the desired degree, under our control, or that the differences in the skill or frequency or intensity or some other dimension with which the behaviors are exercised rest on innate (genetic) differences among us either as individuals or as members of some subset of humans. These kinds of considerations we may dismiss immediately, at least as to what the facts of the matter are, for those facts do not depend to the slightest degree on what anyone *wishes* humans were like.

But on the factual side, as contrasted with the non-relevant normative side, many philosophers have anyway tended to be attracted to accounts of complicated human skills in terms of *learning* as contrasted with, if I may so put it, *genes*. Just why this is, or has been, the case is a result of very complicated social and intellectual factors. But an important part of the story has to do with the tradition of *empiricism* among analytic and scientifically-minded philosophers, going back to Hume and even earlier, a tradition that is suspicious of anything that seems to be of a sort to be called *innate* knowledge. All knowledge, we are told, comes from sense experience and, possibly, introspection and other awareness of our own mental and bodily states; none is already stamped on us at birth or comes only through normal physiological development. For these philosophers, defenders of any kind of innate knowledge are looked on as confused about the requirements of knowledge or else unwilling to find out just how much can be explained by learning.

Until recently, too, the discipline of music psychology largely reflected this frame of reference with its initial suspicion of any kind of innatist theories. The beginnings of music psychology as an empirical enterprise are to be found near the beginnings of experimental psychology itself, that is, in the mid-nineteenth century in the laboratories of such figures as Carl Stumpf and Hermann Helmholtz, but the person widely known as "the father of music psychology," Carl Seashore (who, as it happens, made his career at my own institution of the University of Iowa), comes on the scene only toward the end of the second decade of the twentieth century. Seashore himself, from whatever specific cultural and intellectual factors but consistent with the emerging school of behaviorism in psychology—understood here as insisting on the predominantly *learned* component in explaining human behavior—was strongly anti-innatist with respect to understanding musical behavior; and his views were to form the atmosphere that prevailed in music psychology—largely then an American discipline, at least as an experimental enterprise—for the next half century.

What happened toward the end of that period was, in a word, *Chomsky*, and, in a few more words, the beginning of cognitive science. By now there are many workers in the field of music psychology who are investigating music from the evolutionary, genetic, and neurological perspectives and in ways that almost necessarily raise questions about the possibility of innate factors that underlie our hearing, understanding, and appreciation of music. Still the most important of these studies, I submit, is Lehrdahl's and Jackendoff's book which dates from 1987. And I propose, for the next several pages, to use this important study as a basis for raising the question of what I called knowledge-*for* music, that is, knowledge, if there is any, that must already be in place in order to explain the human response to music.

In their Preface, Lehrdahl and Jackendoff note that among the first to suggest that the "insights" of Chomsky's transformational-generative linguistics might profitably be applied, in the search for a presumably universal "musical grammar," to the domain of music was the distinguished American composer and conductor, Leonard Bernstein, in lectures he gave at Harvard University in 1973 and published in 1976. In fact, it was these lectures that gave rise to seminars and study groups in the Boston area of which their book was one eventual result.

Like Chomsky with respect to language, Lehrdahl and Jackendoff seek a "competence theory" with respect to music. A competence theory would be one that has as its goal "a formal description of the musical intuitions of a listener who is experienced in a musical idiom." Such musical intuitions are further characterized as "the largely unconscious knowledge that the listener brings to his hearing—a knowledge that enables him to organize and make coherent the surface patterns of pitch, attack, duration, intensity, timbre, and so forth." (p. 3) Furthermore, we are to understand that much of this unconscious knowledge—not all—is universal in the species, and as such may plausibly be regarded as *innate*—innateness being usually the best explanation of universality. Lehrdahl and Jackendoff make clear that they are not dealing with "affect," that is, with the emotional side of music, as I shall do in the latter part of this paper, but only with the *structure* that the listener hears in, or attributes to, the music.

Before we turn to some specifics of the theory, we should note that Lehrdahl's and Jackendoff's general conclusions are, indeed, that the understanding of music is "a domain of human cognitive capacity," and that "Much of the complexity of musical intuition is not learned, but is given by the inherent organization of the mind, itself determined by the human genetic inheritance." (p. 281) Thus we are to understand that we human beings, from whatever evolutionary pressures and adaptations, are genetically structured in ways that predispose or constrain or even necessitate us to hear music in certain ways and not in any of the other, numerous or even infinite, merely possible ways. And it is, so we may reasonably assume, in the attribution of structure that lies the difference between the human response to music and the responses of members of other species, many of whom obviously are affected by sounds of the sort we humans call music.

Like Chomsky in his transformational-generative grammar, Lehrdahl and Jackendoff propose to state the heart of their theory as consisting of certain "non-introspectible" *rules*, where a rule expresses "a generalization about the organization that the listener attributes to the music he hears." (p. xii) These rules, on the theory, are of three sorts: first, the *well-formedness rules*, "which specify the possible structural descriptions;" second, the *preference rules*, "which designate out of the possible structural descriptions those that correspond to experienced listeners' hearings of any particular piece;" (p. 9) and third, the *transformation rules*, which "apply certain distortions to the otherwise strictly hierarchical structures provided by the well-formedness rules." (p. 11) The transformation rules on their theory, and unlike the linguistic theories that inspire it, play a very minor role, and so we shall dismiss the possible suspicion that, as characterized, such "rules" are there merely to accommodate those cases that don't fit the theory as defined by the well-formedness and preference rules.

Each kind of rule, in turn, applies to four different aspects of structure; namely, in their terminology, (1) grouping structure, (2) metrical structure, (3) time-span reduction, and (4) prolongational reduction. We need not bother with the precise definitions of these notions to proceed. Here is an example of a well-formedness rule about grouping, their first one:

> Any contiguous sequence of pitch-events, drum beats, or the like can constitute a group, and only contiguous sequences can constitute a group. (p. 37)

If this sounds more like a *definition* of 'group' than a rule about how we hear music (although it does, importantly, say that anything we hear as a unit of music requires contiguity), consider the rather more specific well-formedness grouping rule that:

> If a group G1 contains part of a group G2, it must contain all of G2. (p. 38)

This rule tells us, in effect, that there are no overlapping groups—as defined, I would say. Now here are some examples of preference rules (I've picked some of the simpler ones so we can grasp them more easily, although some of the more complex ones have, as one would expect, more content of substance to them). The first, a grouping preference rule, is:

> Prefer grouping analyses that most closely approach the ideal subdivision of groups into two parts of equal length. (p. 49)

The second, a metrical preference rule and one that is specifically cited as universal by Lehrdahl and Jackendoff, is:

> Prefer a metrical structure in which beats of level *Li* that are stressed are strong beats of *Li*. (p. 79)

These few examples, even when combined with what I have already said about their work, give very little idea of the depth and richness of Lehrdahl's and Jackendoff's theory and its elaboration in their book. But it is enough, I believe, for me to proceed to consider the issues of musical knowledge in the sense of knowledge-for music, knowledge that we have, if any, that we bring to the ways in which we hear music.

We may begin by observing, in presumed agreement with Lehrdahl and Jackendoff, that the experiences and reports of the listener, considered individually and from the point of view of the listener, are of almost no value in assessing their theories of innateness and universality. For just as in the case of language, the individual practitioner would rarely be in a position—and certainly not by virtue simply of knowing a language—to know how the structure of his or her language compares with the range of all abstractly possible structures nor, even more to the point, that the structure of one's own language is, within the probably infinite domain of possible structures, very similar to the structure of all other human natural languages. Similarly, in the case of music, the listener—engaging in whatever reflection and introspection of one's own musical experience as one may—will never discover how the structures that one hears in, or attributes to, the music compare to all possible structures nor, again, how similar the structures one does hear in the music are to the structures most or all other humans do or would hear in the music. No, these matters are not ones to be settled by the report of any individual listener, for they are not matters about the individual listener in the relevant sense—as we have just seen. And I will say also, without further ado but without yet answering the question of whether or not it is *knowledge* that we are dealing with, that we, in fact, do now have more than adequate empirical evidence reasonably to conclude that there are in us innate and therefore universal structures of a specificity that strongly constrain how we hear music.

But are these structures pieces of knowledge, as Lehrdahl and Jackendoff maintain as, for example, in my earlier quote from them? Perhaps more important than arriving at a simple yes or no answer to this question is showing in what ways these innate structures do or may and in what ways do not or may not resemble what philosophers conceive knowledge as being. No doubt some exercises of this sort have already been undertaken with respect to analogous claims about knowledge—including those of Chomsky himself—with respect to language. But it will be useful, I judge, to proceed with the case of music—as may not have been done before.

In the spirit of Gilbert Ryle's most famous chapter of his *The Concept of Mind*, we may begin with the distinction between knowing-*how* and knowing-*that*. Although, as critics have shown, the distinction is not really as precise as it initially appears, we may agree with Ryle's main point in making the distinction; namely, that knowledge-how does *not* presuppose knowledge-that. Knowing how to swim is not the application of theory to practice, even if, in some cases, some knowledge of truths—that is, knowledge-that—may be useful. And I think it is clear that if we agree with Lehrdahl and Jackendoff, as I am prepared to do, that the sort of competences they are talking about can usefully be characterized as the unconscious application of certain rules to the music one hears, then the sort of knowledge that is involved is *at least*—but not necessarily only—a kind of knowledge-how. For being able to apply a rule—and especially a rule of whose content one is not consciously aware and cannot become consciously aware of in oneself—is just to have a certain *capacity* or *ability* or *power*. It is universal among the innumerable capacities that we humans have not just in its subject matter but also in what we might call its *prescriptive* nature—that we *must* hear music in certain ways if we hear it at all. But being constrained or even necessitated to do something presupposes being able to do it, and capacities are just what we are talking about.

But is that all there is to it? That is, granted that the existence of these innate mental structures that apply to the listening of music amounts at the least to possessing certain capacities—a knowing-how to do something—does it amount to that also at the most? In the case of language, Chomsky has long insisted that his findings have constituted a refutation of empiricism and support for rationalism—the doctrine of innate ideas of the early modern philosophers. That, as members of a certain species, we are born with certain capacities of behavior or capacities to acquire capacities never has been, or should have been, in the least degree questionable, for the very notion that an organism (or anything, for that matter) is of a certain natural *kind* clearly implies just that kind of fact for the members of each species. No, those who, in the cases of both language and music, speak of unconscious knowledge and the like mean to say that such knowledge consists, at least in part, of knowledge-*that*, innate structures whose perspicuous representation in language is in the form of *propositions*, that is, entities with *truthvalues* and that can be *believed.* Lehrdahl's and Jackendoff's rules take the form sometimes of categorical and conditional imperatives (in the purely grammatical sense) and also sometimes of categorical and conditional indicatives (which larger distinction, however, does not map directly on to the wellformedness/preference distinction). But whichever form it takes, it is not the content of the rule as such that expresses one's unconscious knowledge but, instead, that there is a rule of such-and-such content that is applicable to this musical situation. Perhaps this distinction between, if I may so put it, the rule as knowledge and knowledge of the applicability of the rule doesn't really matter, and maybe both are supposed to be propositional knowledge; but I am going to proceed on the assumption that what is being claimed is that each of us has, unconsciously and innately, a storehouse of knowings-that that are relevant to how we hear music.

What is it to know *that* something is the case? And how is an answer to that question applicable to the issue of whether or not we have, or could have, innate musical knowledge? The most common answer to the first question, at least among so-called analytic philosophers, is what both its advocates and some of its critics call the *theory* (I would call it either stipulation or anthropological report) that knowledge is *justified, true belief.* We need not enter what to my view is the overblown, self-deceptive, and unfortunately voluminous literature on this topic to extract from it the crucial notion of *justification* for the issues at hand. For empiricism and rationalism were and are theories not only about the sources of, but also the justifications for, our human knowledge.

Assuming then that what Lehrdahl and Jackendoff call unconscious knowledge can be expressed as certain *beliefs* a person has, I think we should grant immediately that there is nothing in the nature of things that prevents the possibility of there being such kinds of beliefs, even if we grant, with one form or aspect of empiricism, that all *simple* ideas come from sense experience and introspection. For if we, perhaps unlike the early modern philosophers, make a sharp distinction between terms and propositions (to speak in one traditional way), we may agree that until we have experience of what is referred to by the terms of the propositions, we don't know what we are talking or thinking about; but that having such experience may then, as it were, "turn on" beliefs embodying those terms, which beliefs then affect subsequent behavior. And that is the idea at work here: the organism behaves as if it believes such-and-such; it doesn't have any conscious beliefs of that kind, so it must have unconscious beliefs of that kind.

And here, perhaps, my skepticism that these innate structures are best considered as consisting in part of beliefs begins to show. Or, perhaps better said, I am inclined to say that the notion of unconscious belief is better restricted, as it was originally, to those situations in which a person's behavior is best explained by the ascription of a belief (or what is more likely to be called a memory, but memories embody beliefs) that is contrary or contradictory to what the person consciously believes. And, as is obvious, this is quite unlike the case of "unconscious knowledge" we may bring to music.

But the main point I want to make, in closing this section of this paper, is that the more important claim of empiricism—that beliefs about the world can be *justified* only by experience of the world—remains untouched by any findings of the sort that could confirm Lehrdahl's and Jackendoff's theories. Perhaps this thesis is more easily seen to be true in the more familiar case of language: if the relevant innate beliefs are to the effect that one's utterances in the language of one's parents, say, must conform to certain structures in order to be recognized, understood, and "approved" by them, then while, on the theory, those beliefs are in fact true, a person could be justified in believing them, even if only unconsciously, by *observations* that the requisite connections hold between structures and understanding. It is, ultimately, itself an empirical question as to how we come to have the beliefs we do have and whether or not any beliefs we have are, as it were, in our genes. But it is a philosophical question as to what, if anything, could make one justified in holding this or that (kind of) belief. And on this matter neither Chomsky nor Lehrdahl and Jackendoff have given us any reason whatsoever to abandon the empiricist's criterion for knowledge and thus the conclusion that, in this strict but important sense of justification, there is not, and could not be, any innate knowledge of music or language or, for that matter, anything at all. This conclusion does not in the least deny the importance of Lehrdahl's and Jackendoff's findings nor those of any other researchers in the field of the innate basis of our musical life. In short, and possibly contrary to the beliefs—maybe unconscious—of some philosophers, the empiricist with respect to justification can be an innatist to whatever degree the findings of empirical science may dictate.

***

There is another kind of knowledge in addition to knowledge-how and knowledge-that; namely, knowledge-*of.* This is the kind of knowledge we gain through *acquaintance* with something—either directly by sense experience and inner awareness or indirectly by having the object represented to us in some way. This kind of knowledge seems obviously irrelevant and indeed almost definitionally so as to the question of knowledge-for music. But, in elaborating on ideas I have put forth in a recent book (1999) in the pursuit of the issue of knowledge-from music, it is exactly a certain sort of knowledge-of with which we shall be concerned. In particular, I want to consider the possibility that we can gain knowledge of certain moods and emotions by having them represented to us by music. We can use Lehrdahl and Jackendoff in order to get into the issue in this way: While acknowledging their debt to Chomsky's theories, they are also eager to stress certain important disanalogies between musical and linguistic theories in maintaining that ". . . whatever music may "mean," it is in no sense comparable to linguistic meaning; there are no musical phenomena comparable to sense and reference in language or to such semantic judgments as synonymy, analyticity, and entailment." (p. 5) While this sentence may be consistent with some theories according to which music represents something extra-musical to the listener, these words with those that follow in their text seem to cast doubt on this possibility or likelihood by emphasizing, correctly, that music is not, in any literal sense, a *language.* But of course, we know that many things that are not language—from flags and paintings, maps and religious icons, gestures and looks, and much, much more—represent and symbolize to the human mind. So the mere fact that music is not language leaves completely open the question of whether or not it represents and, if so, possibly gives us knowledge of that which it does represent.

Probably the best known defender of the theory of music as representation—or what she more often called symbolization—is the American philosopher Susanne Langer. And she was eager also to maintain also that music can represent what language cannot, as the following passages from her best-known book, *Philosophy in a New Key*, indicate:

> . . . it seems peculiarly difficult for our literal minds to grasp the idea that anything can be *known* which cannot be *named.* (p. 232) What is here criticized as a weakness [music's not having certain features of

> language], is really the strength of musical experience: that *music articulates forms which language cannot set forth.* (p. 233) The imagination that responds to music is personal and associative and logical, tinged with affect, tinged with bodily rhythm, tinged with dream, but *concerned* with a wealth of formulations for its wealth of wordless knowledge . . . (p. 244)

Langer's *ineffability* thesis—that of "wordless knowledge"—is not essential to my argument although, as it happens, I do hold a version of it myself. What is essential is the thesis that we can gain knowledge from music of something that is not music, whether or not that knowledge might be gained or represented in other ways. Theories of the sort we have been considering in the first part may well tell us *how* we listen to music while also telling us something about human nature, but they have little or nothing to say about the *why* of the matter. Lehrdahl and Jackendoff said they were not concerned with the affective, that is, the emotional, side of music, and rightly so given their concerns. But as to the question itself, my broad thesis in a nutshell is that we listen to music because doing so *arouses* emotion in us; and it arouses emotion because it *presents* us with emotion (though not necessarily or even usually the same emotion), that is, it represents emotion and thereby gives us knowledge of what some emotional states are. The question is how this representation might be accomplished, through what ground and mechanism.

But let us ease ourselves into this matter by asking yet again why music matters so much to human beings? Why, to speak more precisely in one dimension, is it *sounds* and not visual or kinesthetic or any other kind of sense or intellectual "data" that, at least among artworks, affects humans more widely and more profoundly than any other? Why or how, to speak more precisely in another dimension, have there come to exist beings who are willing to spend so much of their time and resources on music and many of whom are content to sit for hours on end, virtually silent and motionless, for the purpose of listening to, and watching, performances of music?

Surely there must be something interesting about our human nature, as produced by evolution, and something interesting about the nature of sound, at least as it presents itself to awareness, that are relevant to any full answer to these questions. But it is clear that the evolutionary part of the answer, whatever it is, cannot by itself fully answer the question if only because it involves no investigation into the nature of sound. Nor can it be adequate to suggest that we humans just happen to enjoy music; for the empirical absurdity of treating the relation between human beings and music as mere "stimulus/response" is attested to, or at least strongly supported by, numerous facts: among them that most pieces of music involve consciously constructed and perceived structures of a certain sort; that listening to music requires effort and attention and memory of sorts that may reasonably be called *cognitive*; that the pleasures of music are, while occasionally also physical, for the most part of a sort difficult to categorize or characterize but that certainly involve the *appreciation* of certain, often complex properties of the music one is hearing; and that in talking about music, we use the language of the emotions and other aspects of mental life to characterize the music itself.

Langer's theory, according to which music *symbolizes* the emotions by virtue of a shared logical form, I believe to be substantially correct—as far as it goes. But neither Langer nor any previous thinker has explained how it is that sounds might have this symbolic capacity. More generally, no one yet has embodied this idea in theories of the nature of representation, the nature of sound, and the nature of consciousness in a way that would explain just why it is sounds and not other phenomena that affect us as music does.

My one-sentence version of the general theory is this: *Passages of music are quasi-natural representations of emotions and other states of consciousness.* Consciousness figures in this theory, as in Langer's, *both* as what *is* represented *by* music and as what is represented *to*, by music. Thus a theory of the nature of consciousness would seem especially called for by any such account. And because it is sounds that are doing the representing, some theory of their nature would seem also to be relevant. But on this occasion, and in keeping with the emphasis on cognitive science, I shall focus on the nature of representation, having, in the end, to take certain assumptions about consciousness and sound for granted.

Quasi-natural representation, as I call it, is to be distinguished on the one side from purely conventional representation in which it is entirely a matter of choice (usually the "choice" of the linguistic community) what represents what, and on the other side from purely natural representation in which the representing entity, by its inherent nature, points to what it represents. Language is the best example of conventional representation and consciousness itself (or some constituent of it) is probably the only example of natural representation. And while natural representation is the fundamental form of representation as that on which, in different ways, all other forms depend, quasi-natural representation is the most complex and, especially if it exists, probably the most interesting form of representation.

Considering only the human case, we may say that there is quasi-natural representation if, *due only to the natures of the representation and of being human*, some kind of entities do represent something in particular to human beings. Thus, this is a kind of representation that is beyond choice to change although, as always, what does represent in this way, if anything, can always represent in the conventional way by being stipulated to represent something or other. In short, in quasi-natural representation, the fact that something represents something else to a human being is a *law of nature* (or an instance of a law). It might well be argued that human nature is always involved in representation as well as the nature of the representation at least to the extent that the representation, if it is to be such in any systematic way, must be of a sort that is easily perceivable by human beings, something that we can easily manipulate, and so on. I have no objection to the idea that there is a continuum and not a sharp line between conventional representation and quasi-natural representation; for it is surely the case, as Chomsky has impressed on us, that, due to human nature, there are limits at least on the forms that language can or will take, to mention only the most important case. And it will probably not be doubted that the fact that humans have such extensive conventional systems of representation at all is due to, and importantly expressive of, our human nature.

Even if it is a truism that all known representation depends, in some way, on the nature of the human mind, the distinction between conventional and quasi-natural representation is not thereby rendered vacuous. To see this more clearly, let us understand that in the strict sense quasi-natural representation must involve the possibility of being represented to whether or not the person is consciously aware of the fact of such representation. This fact is for many a reason to dismiss the notion immediately as either logically or empirically absurd. For, as defined, if someone is perceiving something that by its nature represents something to any human being by our nature, that person is, *ipso facto*, being represented to whether or not that person knows or suspects that representation. What this could possibly mean exactly remains to be explained, nor can I do so on this occasion. But it may help to suggest that the best-known case of what may be a kind of quasi-natural representation, that of dreams, would seem to *require* just this perplexing phenomenon of being represented to without conscious awareness of it. And I do mean by that not just that the person had the sensory awareness of something that sometimes brings something to mind in someone but that that something was actually brought to mind in that person even though he or she was not consciously aware of that fact. In any case, this feature distinguishes the notion of quasi-natural representation from representation in language in which some analogous phenomenon *may* occur but is not required by the very notion of language itself. And it will be well to keep in mind that the evidence that a person has been represented to even while not knowing or suspecting or even while denying such a representation is that it is the *best explanation* of that person's actual and dispositional behavior.

Quasi-natural representation, like conventional representation, is a minimally *three-term* relation involving the representation itself, the represented, and the person represented to. Whatever isomorphism or resemblance may exist between the entities that are the representation and "entities" (they may not exist) that are represented, there is no simple or natural, that is, *two-term* relation between the former and the latter. The one represents the other only *to* or *for* someone in some way *because* of the human mind—the mind as *intermediary*, in a proper but broad use of that word—although here because of its *nature* and not its *will* or *choice.*

How is it, then, that *sounds* and perhaps sounds alone, can be, for human beings, quasi-natural signs of certain states of consciousness such as the having of moods and emotions? I suggest that *part* of the answer to that question—here to be only asserted and not argued—is that sounds and awarenesses have a certain *ontological affinity.*

But to see briefly what that might mean, consider, first, that it seems to be the case that, *alone among physical phenomena known to us, sounds require duration but do not require change.* Or, to speak a little more precisely: in the physical world, the only temporal property that does not require change is that of *being-a-sound* and whatever "other" properties that involves such as pitch, timbre, and volume. The exemplification of any other physical temporal property such as *being-a-birth* or *being-a-revolution* requires not only duration but change as well. This gives sound a very special ontological status in the physical realm, but I don't know how to prove that it is so beyond asking you first to consider the nature of the property of *being-a-sound* and then to try to think of some other physical property that shares this feature with that property, believing, of course, that you cannot succeed in the latter.

If this ontological feature of requiring time but not change is unique to sound in the physical realm, it is not unique in reality as a whole. Or, at any rate, I now submit that states of consciousness are also such that they require time but not change. This feature of consciousness states (which, following the likes of Husserl and Sartre, I take to be the same as *intentional* states) I have argued for in detail in two books (1989, 1999), but I am by no means the first to suggest that there is a profound connection between consciousness and time. At any rate, here I must simply assume this feature of conscious states as established.

We have thus come upon a profound *ontological affinity* between consciousness and sound: *states of consciousness and sounds alone are such that, while they require time for their existence, they do not require change.* This affinity that sets them apart from everything else in this crucial regard suggests that, so to speak, consciousness might find in sound an "image" of itself, not just in the feature that I have identified but in more complicated ways as well. This is the ontological basis of my thesis that passages of music are quasi-natural representations of possible states of consciousness.

What I cannot do on this occasion is to argue, instead of merely claim, that the theory that music in fact *represents* the emotions to us is the *best explanation* of why music *arouses* emotion in us, and also of why, what is quite a different fact, we spontaneously use the language of emotion in talking about the music itself. Nor, alas, can I show, instead of merely assert, the intelligibility of supposing that much of this representation takes place through unconscious awareness where by 'unconscious awareness' I mean occurrent states of consciousness which the person does not recognize him- or herself as having and may even deny having. There is surely no serious doubt about the existence of the phenomenon of unconscious awareness but only in what kinds of situations it takes place and in what modes and with what objects. I believe we will have a better understanding of the importance that music has for our species if we take seriously the possibility of both the conscious and the unconscious awareness of things that are not music that are, as it were, forced on us by listening to music.

Finally, in suggesting that parts of the mind that are not part of our fully conscious existence play important roles in both what we bring to music and what we get from music, I hope slightly to have advanced the understanding, empirical and philosophical, of how it is that music has such power and mystery for human beings.

REFERENCES

Addis, Laird (1989). *Natural Signs: A Theory of Intentionality.* Philadelphia: Temple University Press.

Addis, Laird (1999). *Of Mind and Music.* Ithaca: Cornell University Press.

Bernstein, Leonard (1976). *The Unanswered Question: Six Talks at Harvard.* Cambridge: Harvard University Press.

Chomsky, Noam (1966). *Cartesian Linguistics: A Chapter in the History of Rationalist Thought.* New York: Harper & Rowe.

Hodges, Donald, ed. (1996). *Handbook of Music Psychology* (2nd edition). San Antonio: IMR Press.

Langer, Susanne (1942). *Philosophy in a New Key: A Study in the Symbolism of Reason, Rite, and Art.* Cambridge: Harvard University Press.

Lehrdahl, Fred and Jackendoff, Ray (1987). *A Generative Theory of Tonal Music.* Cambridge: The MIT Press.

Raffman, Diana (1993). *Language, Music, and Mind.* Cambridge: The MIT Press.

Ryle, Gilbert (1949). *The Concept of Mind.* London: Hutchinson's University Library.

Chapter 10

EVENT COREFERENCE AND DISCOURSE RELATIONS

Laurence Danlos & Bertrand Gaiffe
UFRL-Université Paris 7 & LORIA-CNRS, Vandoeuvre lès Nancy

Abstract: Event coreference is largely ignored in works on coreference as well as in works on temporal order in discourses, in which temporal relations between two eventualities e_1 and e_2 always suppose that $e_1 \neq e_2$. However, event coreference is not an uncommon phenomenon. We show that it is the keystone on which at least four types of discourses are based: on the one hand, particularizing and generalizing discourses, which can be viewed as special cases of elaboration and reformulation discourses respectively, on the other hand causal discourses in which the effect is expressed by means of a causative verb, which are generally analyzed as explanation or result discourses. Descriptions and linguistic analyses of these discourses allow us to present unusual linguistic phenomena (e.g., coreference between existentially quantified elements, or assymetrical behavior of explanation or result discourses). The discourse relations involved are examined. This leads us to introduce and define new discourse relations. It also sheds a new light on the discourse relations Explanation and Result.

1. INTRODUCTION

Works on temporal relations between two eventualities e_1 and e_2 examine almost exclusively if one eventuality precedes, includes or overlaps with the other one (Moens & Steedman 1988, Asher 1993, Pustejovsky 1995). All these temporal relations suppose that $e_1 \neq e_2$. We will concentrate on cases where $e_1 = e_2$, i.e., on *event coreference*. Event coreference can be seen as a particular case of either event elaboration or event reformulation, for which

K. Korta and J. M. Larrazabal (eds.), Truth, Rationality, Cognition, and Music, 191-209.

it has been acknowledged that time is not understood as moving (Spejewski & Carlson 1991).

Contrary to coreference between objects, coreference between events has hardly been studied in detail, except for a (pro)nominal phrase referring to an event as in (1) in which the pronoun *this* refers to the event described in the first sentence, see among others Webber (1988), Asher (1993).

(1) Ted arrived late. This irritated Mary.

Nevertheless, there is a number of other cases of event coreference since *the description of an event e can be of three types*: it is either a sentence, or a (pro)nominal phrase or an abstraction which is not linguistically realized. This last case occurs when e is a sub-event of a complex event (cf. infra). Between two successive descriptions D_1 and D_2 of the same event e, there exist therefore $3^2 = 9$ cases of event coreference potentially observable depending on the types of D_1 and D_2. In this paper, the following cases are examined:

- In Section 2, event coreference when D_1 and D_2 are both sentences (noted as S_i) as in (2a) and (2b) is examined.

  (2) a. The yugu team won a victory. They beat the tara team yesterday.
  b. The yugu team beat the tara team yesterday. Therefore, they won a victory.

  In (2a), the two sentences refer to the same event, with more information on that event in the second sentence (which "particularizes" the first one). In (2b), obtained from (2b) by reversing the order of the sentences, S_2 "generalizes" S_1.

- In Section 3, event coreference when D_i ($i = 1, 2$) is a sentence and D_j ($j = 1, 2, j \neq i$) is not syntactically realized is examined. More precisely, *causal discourses* in which the effect is expressed by means of a causative verb (e.g., *crack*) are examined, as in (3a) and (3b) ((3b) is obtained from (3a) by reversing the order of the sentences).

  (3) a. Ted cracked the carafe. He hit it against the sink.
  b. Ted hit the carafe against the sink. He cracked it.

  The causing sub-event of the complex event described by *crack* is not linguistically realized, it is just known that it is an act by Ted on the carafe. However, it is understood as coreferent to the event described in the cause sentence (*Ted hit the carafe against the sink.*).

Studying event coreference requires the definition of *two types of coreference relation according to the quantity of information conveyed*:

Two successive descriptions D_1 and D_2 of the same entity x (event or object) are in a ***particularization*** *relation noted as $D_2 = PART(D_1)$ iff D_2 conveys some new information about x when compared to the information known from D_1.*

Two successive descriptions D_1 and D_2 of the same entity x (event or object) are in a ***generalization*** *relation noted as $D_2 = GEN(D_1)$ iff D_2 does not bring any new information about x (and $D_1 \neq D_2$).*

Let us illustrate these two types of coreference with the well-known case where x is an object. D_1 and D_2 are then (pro)nominal phrases. In (4a), the demonstrative NP *this New Yorker girl* = PART(*an actress*) since it brings new information (*the actress is a New Yorker girl*). In (4b), the pronoun *she* = GEN(*an actress*): a pronoun cannot bring any new information. Finally, in (4c) the definite or demonstrative NP *the / this artist* = GEN(*an actress*) since *artist*, hyperonym of *actress*, does not bring new information.

(4) a. Ted had a drink with an actress. This New Yorker girl irritates Mary.
b. Ted had a drink with an actress. She irritates Mary.
c. Ted had a drink with an actress. The / This artist irritates Mary.

These two types of coreference relation according to the quantity of information conveyed do not play a major role in the literature on object coreference. However, they are essential for event coreference. In the next sections, it will be shown that (2a) and (3a) involve an event coreference relation of type particularization, (2b) and (3b) an event coreference relation of type generalization. This difference in the type of coreference explains assymetrical data. For example, it will be shown that any kind of adjunct can be inserted in the first sentence of (2a) or (3a), while the insertion of adjuncts in the first sentence of (2b) or (3b) is limited. As a consequence, the causal discourses in (3a) and (3b) are not really converse, despite what is generally believed. More generally, it will be shown that particularization and generalization are not symmetrical: the inverse of a generalization relation is a particularization relation, but the inverse of a particularization relation is not always a generalization relation.

We will show that it is necessary to lay down *new discourse relations* (the name of a discourse relation starts with a capital letter): Particularization (a particular case of Elaboration) for (2a) and Generalization (a particular case of Reformulation) for (2b). Of course, these new discourse relations are closely related to the types of event coreference: particularization (abbreviated as PART) for (2a) and generalization (abbreviated as GEN) for (2b). Moreover, we will show that these discourse relations are involved in causal discourses: Particularization for (3a) and Generalization for (3b). We will question the analysis of (3a) with Explanation and that of (3b) with Result, and shed a new light on these causal discourse relations. Discourse relations will be studied in the formal framework of SDRT (Segmented Discourse Representation Theory, Asher 1993, Asher & Lascarides 1995, Asher & Pustejovsky, 2000, Asher & Lascarides 2003).[1]

2. EVENT COREFERENCE BETWEEN TWO SENTENCES

Sections 2.1 and 2.2 provide linguistic descriptions, which are formalized in Section 2.3. Section 2.4 examines the discourse relations involved.

Notation: The symbol X^r_j with i = 1 or 2 denotes the predicate or an argument or modifier in S_i with the role r; for an argument r = agent, patient,... ; for a modifier r = time, location,... ; for the predicate, r is irrelevant.

2.1 Particularizing discourses

A particularization event coreference relation as defined above is to be found in "particularizing discourses" (henceforth PDs) such as (2a) or (5).

(5) Fred damaged a garment. He stained a shirt.

In (5), the same event is described in both sentences, with more information in S_2. S_2 particularizes S_1 since its predicate is a hyponym of the predicate in S_1 (*stain* < *damage*) and the (indefinite) object is a hyponym of the (indefinite) object in S_1 (*shirt* < *garment*), while its subject is an (pronominal) anaphora of the subject in S_1.[2]

More natural examples of PDs are observed when extralinguistic knowledge is at stake, as in (6). These PDs are based on an extension of hyperonymy, namely *X compliment Y > X tell Y that Y is pretty* in (6a), and *X be sick > X have the flu* in (6b). These extended hyperonymy relations rely on cultural or encyclopedic knowledge.

[1] We assume that the reader knows SDRT. If it is not the case, she can skip the SDRT formalizations presented here.

[2] The two sentences are in the same tense, which is required for event coreference.

(6) a. Fred complimented Sue. He told her that she is pretty.
b. Fred is sick. He has the flu.

Our linguistic study is not corpus based. Nevertheless it is easy to find real examples of PDs, such as (7a) found in (Asher 1993, p. 1).

(7) a. (...) linguists and philosophers (...) developed a sophisticated understanding of events and states and the expressions that denote them. They developed a typology of such entities.

The particularization interpretation of a PD can be explicitly marked, as in (8) with the cue phrase *more precisely*.

(8) Fred damaged a garment. More precisely, he stained a shirt.

The PD in (5) or (8) gives rise to an unusual phenomenon: the two indefinite NPs *a garment* and *a shirt* are coreferent. This is unusual since it is generally admitted (Kamp & Reyle 1993, Corblin 1995) that an indefinite NP has an existential reading (i.e. introduces a new discourse referent) and that therefore two indefinite NPs are not coreferent.[3] This is the case in (9) in which two garments are involved, whereas only one garment is involved in (5). We will see in Section 2.4 how to compute the coreference of two indefinite NPs in PDs.

(9) Fred damaged a garment. Joe tore a shirt.

A discourse such as (5) in which at least one element in S_2 specifies a corresponding element in S_1 is called a "PD by specification". Another type of PD is given in (10). In (10), the modifier *during dinner* in S_2 provides temporal information about e; the other elements X^r_2 in S_2 anaphorize or repeat the corresponding elements X^r_1 in S_1.

(10) Fred stained a shirt. He stained it during dinner.

A discourse such as (10) in which at least one element X^r_2 in S_2 has no corresponding element X^r_1 in S_1 is called a "PD by adjunction".

[3] A few cases exist in which two indefinite NPs are coreferent, (ia) and (ib) taken from Corblin 1994.
(i) a. A man arrived. He was a New Yorker.
b. A man, a New Yorker, arrived.
However, the examples presented here, with two coreferent indefinite NPs, have not been described in the literature, as far as we know.

Specification and adjunction may be combined in a single PD. In (11), *a shirt* specifies *a garment* and the time adjunct in S_2 has no corresponding element in S_1.

(11) Fred stained a garment. He stained a shirt during dinner.

At face value, one gets a PD when each element X^r_2 in S_2 is either an (extended) hyponym, an anaphora or a repetition of a corresponding element X^r_1 in S_1 or has no corresponding element X^r_1 in S_1.

To summarise, in text understanding, it has to be computed that there is only one garment and one event involved in (5).[4] In text generation, it must be determined under which conditions a PD may be uttered and which linguistic forms should be produced (Roussarie 2000).

2.2 Generalizing restatement discourses

A generalization event coreference relation as defined in Section 1 is to be found in "generalizing restatement discourses" (henceforth GDs) such as (12).

(12) a. Fred stained a shirt. Therefore, he damaged a garment.
b. Fred murdered Sue. Therefore, he committed a crime.

While a PD stands mainly at the informational level, a GD stands mainly at the intentional level: the first sentence in (12a) or (12b) describes what the speaker wants the hearer to know about an event e, the second one consists in her restatement of e. By definition, a restatement cannot bring new information, but it can present an event in a new light, for example, if the speaker has the intention to forge links with other data, (13).

(13) Fred murdered Sue. Therefore, he committed a crime for which he risks the death sentence.

(12a) is created from (5) by reversing the order of the sentences (and the anaphoric relations), and by introducing a cue phrase like *therefore*, with an epistemic value (Rossari & Jayez 1996). S_2 generalizes S_1 because *damage* and *garment* are respectively hyperonyms of the corresponding elements in S_1. In (12b), S_2 generalizes S_1 because the complex predicate *commit a crime* is a hyperonym of *murder*.

[4] See also Humphreys et al. (1997) for event coreference issues within an information extraction system.

As for PDs, more natural examples of GDs are observed when an extended notion of hyperonymy is at stake, (14).

(14) a. Fred told Mary that she is pretty. Therefore, he complimented her.
b. Fred has the flu. Therefore, he is sick.

At face value, one gets a GD when each element X^r_2 in S_2 is an (extended) hyperonym, an anaphora or a repetition of a corresponding element X^r_1 in S_1.

2.3 Formalized conditions

In (Danlos 2001), the conditions for PDs or GDs have been formalized in the framework of SDRT, using a notion of implication between two DRSs. Intuitively, in first approximation, in a PD, S_2 "implies" S_1. Informally, staining a shirt "implies" damaging a garment, so *Fred stained a shirt* "implies" *Fred damaged a garment* in (5). At the opposite, in a GD, S_1 "implies" S_2.

Nevertheless, the situation is more complex: it is indeed the case that S_1 always "implies" S_2 in a GD. However, it is not always the case that S_2 "implies" S_1 in a PD. There is an assymetry between PDs and GDs. More precisely, *if S_1. (therefore) S_2.* is a GD, then *S_2. S_1.* is a PD. But the converse does not hold, as illustrated in (15). (15a) is a PD. S_1 in (15a) includes a time adjunct (*during dinner*) which is not repeated in S_2, although it is understood that the staining event occurred during dinner and that it is the same event as the damaging event. This means that the second sentence of a PD has to bring new information but has not to repeat all the information conveyed in the first sentence. On the other hand, the second sentence of a GD cannot bring new information at all. (15b) is incoherent (hence the sign #) and cannot be interpreted as a GD.

(15) a. Ted damaged a garment during dinner. He stained a shirt.
b. # Ted stained a shirt. Therefore, he damaged a garment during dinner.

Another type of PD examples with no corresponding GD is illustrated in (16a): (16a) is a PD although the verb in S_2 is a *hyperonym* of the verb in S_1. It has no corresponding GD —(16b) is not a GD. (16a) is paraphrased by (16c) with the anaphoric form *do it*.

(16) a. Ted stained a shirt. He damaged the garment during dinner.
b. # Ted damaged a shirt during dinner. Therefore, he stained it.
c. Ted stained a shirt. He did it during dinner.

To put it in a nutshell, the conditions to observe a GD (noted C_{GD}) are based on a notion of implication between the DRSs which represent the sentences, while the conditions to observe a PD (noted C_{PD}) are slightly more complex.

2.4 Discourse relations

It is likely that anyone working on discourse relations would lay down that the discourse relation between the two sentences of a PD is Elaboration. However, we are going to show that it is necessary to postulate the existence of a new discourse relation called "Particularization" which implies an event coreference. Beforehand, a terminological remark: "particularization" (abbreviated as PART) designates a type of coreference as defined in section 1; "Particularization" designates a discourse relation between two sentences. These two notions are linked in SDRT in the following way: if π_i is the label for the DRS K_i representing the sentence S_i, then $K_2 = PART(K_1) \Leftrightarrow$ Particularization(π_1, π_2).

Elaboration is given various definitions (Hobbs 1979, Mann & Thomson 1988, Asher & Lascarides 1995), but whatever its definition, this discourse relation implies no event coreference relation. Let us illustrate this point with a "classic" example of Elaboration, (17a) or its shorter version, (17b). The first sentence denotes an event e_1 which is "elaborated" in the next sentence(s) which bring(s) new information on **one** or **some** stages (sub-events) of e_1. No event coreference is at stake: a trip by plane cannot not be reduced to a takeoff and a landing and even less to a simple takeoff. There exists a number of stages in the trip which are described neither in (17a) nor in (17b).

(17) a. Nicholas flew from Austin to Paris. He took off at 6 am. He landed at 2 pm.
b. Nicholas flew from Austin to Paris. He took off at 6 am.

Therefore it is necessary to postulate the existence of a discourse relation (Particularization) which implies an event coreference relation.[5] Only this discourse relation allows us to preserve well established insights on

[5] In the framework of SDRT, the indefeasable consequence of Particularization is written in the following way: Particularization(π_1, π_2) $\rightarrow$ Main-event(π_1) = Main-event(π_2).

discourse referents. The argumentation is the following: examples such as (5) are productive counter-examples to the principle of the existential reading of an indefinite NP (which entails that two indefinite NPs are not coreferent). Hence, two solutions towards this principle are available:

- Given these productive counter-examples, this principle is abandoned and an anaphoric reading of indefinite NPs is proposed to allow the coreference of two indefinite NPs. This solution does not seem sound since this principle is nearly always true, except in PDs (and in GDs and in some well known exceptions, see note 2).
- This principle is not abandoned. The two indefinite NPs in (5) are represented in the standard way: two discourse referents x and y with the complete conditions garment(x) and shirt(y) are introduced. Yet the coreference relation $x = y$ is established thanks to the discourse relation Particularization. This discourse relation implies an event coreference relation which implies in turn that the arguments with the same role are coreferent. In other words, it is because the discourse relation Particularization is established between the two sentences of (5) that it is known that the two sentences refer to the same event, and therefore that the two patients *a garment* and *a shirt* are coreferent, i.e., $x = y$.

This last solution seems better. It raises the following question: how can it be established that the discourse relation in (5) is Particularization (which amounts to establishing that (5) is a PD)? The answer to this question lies in linguistic knowledge: the two sentences in a PD follow some strong linguistic constraints which have been described in Sections 2.1 and 2.3. This set of constraints, noted as C_{PD}, allows the computation of the discourse relation Particularization. In SDRT, the rule is the following (where > is the symbol for the default implication): $(\tau, \pi_1, \pi_2) \wedge C_{PD} > \text{Particularization}(\pi_1, \pi_2)$.

Summary on event coreference between two sentences

It is necessary to postulate the existence of a new discourse relation, Particularization, which implies an event coreference relation between the events e_1 and e_2 described in the first and second sentences. It allows the establishment of an event coreference relation although the events involved are existentially quantified. This event coreference relation implies the coreference of participants with the same thematic role, even if they are existentially quantified. This discourse relation differs from Elaboration for which only one or some stages (sub-events) of e_1 are specified. Nevertheless, Particularization can be seen as a particular case of Elaboration. For exactly the same reason, a discourse relation Generalization has to be laid down in order to establish the event co-reference in such examples as (12).

3. CAUSAL DISCOURSES

3.1 Linguistic description

Examples (3a) and (3b), repeated in (18a) and (18b), express a "direct causation" (Fodor 1970, Schank 1975). It is generally admitted that the discourse relation is Explanation in (18a) and Result in (18b). We adopt this position in this section before questioning it in the next section.

(18) a. Ted cracked the carafe. He hit it against the sink.
 b. Ted hit the carafe against the sink. He cracked it.

In these causal discourses, the effect is expressed by means of a causative verb, here *crack*. It is well known (see among others Moens & Steedman 1988, Pustejovsky 1991, Pustejovsky 1995) that a causative verb denotes a complex event which can be broken down into:

- a causing sub-event, here a non specified[6] act by Ted on the carafe which is represented in the simplified[7] DRS π_1: e_1–?ACT(t, c),
- and a resulting state, here the cracked state of the carafe which is represented by π_2: e_2– cracked(c).

π_1 and π_2 are grouped in the DRS π_3, which is noted as $\pi_3 = (\pi_1, \pi_2)$. π_3 can be viewed as the DRS representing the sentence with the causative verb (*Ted cracked the carafe.*). The analysis of a clause with a causative verb is sketched in Figure 10-1.

Ted cracked the carafe

π_1= e_1Š?ACT(t, c)
π_2= e_2Š cracked(c)
π_3= (π_1, π_2)

Figure 10-1. Analysis of a clause with a causative verb

[6] The act by Ted on the carafe is not specified, however it is understood as an act which affects its patient, i.e., the carafe.

[7] Our simplified graphic representation of DRS includes conditions (in a compact way) but not the universe. We leave it to the reader to re-establish the complete DRSs.

The causal interpretation of discourses such as (18a) or (18b) can be explained in linguistic terms thanks to the notion of event coreference.

The cause sentence denotes an event which is interpreted as coreferent to the causing sub-event of the complex event referred to by the causative verb.

For example, in (18a) or (18b), the hitting event is coreferent to the causing sub-event of *crack*, i.e. the non specified act by Ted on the carafe. Let us underline that the causal interpretation of the discourses we are concerned with is totally bound to event coreference. To emphasize this point, consider the discourses in (19). Normally they should not be given a causal interpretation. However, if the hearer imagines that Ted is James Bond equipped with a special camera, then she may assume a causal interpretation of (19). In such a case, she assumes an event coreference between the *taking a photo* event[8] and the causing sub-event of *crack*.

(19) a. Ted cracked the carafe. He took a photo of it.
b. Ted took a photo of the carafe. He cracked it.

The analysis of (18a) is schematized in Figure 10-2 with the event coreference relation $e_1 = e_4$, where e_4 is the hitting event. π_1 introduces an underspecified predicate ?-ACT. This predicate is specified coreferentially in π_4 which labels K_4, the DRS for the hitting sentence. *hit* can be considered as an hyponym of ?-ACT. The event coreference relation $e_1 = e_4$ is therefore of type particularization, i.e. K_4 = PART(K_1), since K_4 brings new information as compared to K_1. Thus, Particularization (π_1, π_4) can be laid down.

The analysis of (18b), similar to that of (18a), is schematized in Figure 10-3 with the event coreference relation $e_5 = e_6$.[9] This coreference relation is of type generalization, i.e. π_6 = GEN(π_5), since K_6 does not bring new information compared to K_5. K_6 can be viewed as an anaphora by hyperonymy of K_5.

[8] The taking a photo event is then understood as an act which affects its patient, see note 6.

[9] The numbering π_5-π_8 is used instead of the standard one π_1-π_4 to refer later to the DRSs in figures 1 and 2 without ambiguity.

Ted cracked the carafe. He hit against the sink	*Ted hit the carafe against the sink. He cracked it.*
π_1= e_1Š?ACT(t, c) π_2= e_2Š cracked(c) π_3= (π_1, π_2) π_4= e_4Š hit(t, c, s)	π_5= e_5Šhit(t, c, s) π_6= e_6Š?ACT(t, c) π_7= e_7Šcracked(c) π_8= (π_6, π_7)
K_4= PART(K_1) $\Leftrightarrow$ Particularization(π_1, π_4) with $e_1 = e_4$	K_6= GEN(K_5) $\Leftrightarrow$ Generalization(π_5, π_6) with $e_5 = e_6$

Figure 10-2. Analysis of (18a) (preliminary version) *Figure 10-3.* Analysis of (18b) (preliminary version)

The undebatable link between causal interpretation and event coreference is shared by both Explanation and Result discourses. However these discourses differ in that Explanation discourses rely on an event coreference of type particularization, whereas Result discourses rely on a generalization event coreference. In parallel with the fact that some PDs do not have a corresponding GD (Section 2.3), some Explanation discourses do not have a corresponding Result discourse. Let us illustrate this point with the adjunction of modifiers in the sentence with a causative verb.[10] Two types of adjuncts must be distinguished:

- those with scope over the causing sub-event, for example, temporal adverbs (*during dinner*). In *Ted cracked the carafe during dinner*, *during dinner* has scope over the causing sub-event. The information known on this event is: e_1–?ACT(t, c), $\wedge$ *during-dinner*(e_1).

- those without scope over the causing sub-event, for example "quantifier" adverbs (*badly*). In *Ted cracked the carafe badly*, *badly* has scope over the resulting state but not over the causing sub-event. The information on the causing sub-event is e_1–?ACT(t, c), and that on the resulting state is e_2–*cracked(c)* $\wedge$ *badly(e_2).*

The adjunction (in the sentence with a causative verb) of a modifier with scope over the causing sub-event, e.g., temporal adverb, does not alter the causal interpretation when Explanation is involved, see (20a) with the adjunction of *during dinner* in the *crack* sentence. But it does alter it when

[10] The claim that Explanation discourses rely on an event coreference of type particularization, whereas Result discourses rely on a generalization event coreference is thoroughly justified in (Danlos 2000).

Result is involved, see (20b) which is marked with the sign # to indicate that no causal interpretation is at stake.[11] This assymetry can be explained by a principle on the type of the event coreference relation: it has to be of type particularization for Explanation, as it is in (20a), see (21a) which indicates the information known from (20a). It has to be of type generalization for Result, which is *not* the case in (20b): see (21b), which indicates the information known from (20b) and where the generalization coreference relation is *negated.*

(20) a. Ted cracked the carafe during dinner. He hit it against the sink.
b. # Ted hit the carafe against the sink. He cracked it during dinner.

(21) a. $e_4\text{-}hit(t, c, s) = PART(e_1\text{-}?ACT(t, c) \wedge during\text{-}dinner(e_1))$ with $e_1=e_4$
b. $\neg[e_6\text{-}?ACT(t, c) \wedge during\text{-}dinner(e_6) = GEN(e_5\text{-}\ hit(t, c, s))]$ with $e_5=e_6$

On the other hand, the adjunction of a of a modifier without scope over the causing sub-event, e.g., "quantifier" adverb, is possible in both Explanation and Result discourses while maintaining a causal interpretation, (22). This goes along with the fact that the coreference relation is of type particularization in (22a), see (23a), and of type generalization in (22b), see (23b).

(22) a. Ted cracked the carafe badly. He hit it against the sink.
b. Ted hit the carafe against the sink. He cracked it badly.

(23) a. $e_4\text{-}hit(t, c, s) = PART(e_1\text{-}?ACT(t, c)$ with $e_1=e_4$
b. $e_6\text{-}?ACT(t, c) = GEN(e_5\text{-}\ hit(t, c, s))]$ with $e_5=e_6$

Let us move to psychological causative verbs (e.g. *irritate*) whose subject can refer to an event (*This incident irritated Mary*) contrarily to causative verbs which denote a physical change of state (**This incident cracked the carafe*). Two types of discourses are thus available to express Result when the change of state is psychological, as illustrated in (24): the subject of *irritate* refers to the cause in (24a) and to the agent of the cause in (24b). Our analysis puts forward that these two discourses, which describe the same

[11] As shown by the fact that it is impossible to insert a causal connective such as *therefore* between the two sentences.

situation but differs in communicative structure, involve both an event coreference relation. This coreference relation is "classic" in (24a): (pro)nominal NP referring to an event. In (24b), it holds between the event referred to in the cause sentence and the causing sub-event of the causative verb. It can be considered that (24b) involves a metonymy: *he* is used instead of *his act*. With a psychological verb, Result can therefore be expressed by the metonymic or full form. With a causative non-psychological verb, only the metonymic form is available.

(21) a. Ted hit the carafe against the sink. This / This incident irritated Mary.
b. Ted hit the carafe against the sink. He irritated Mary.

3.2 Discourse relations

We are going to shed a new light on Explanation and Result. It is common knowledge to state that these discourse relations apply when a causal relation is involved, and when, for Explanation, the description of the effect precedes that of the cause, while for Result, the inverse discursive linear order is observed. These discourse relations can thus be established only when the following axioms (necessary conditions) are verified:[12]

A1: *Explanation and Result hold between two DRSs, one representing the cause, the other one the effect.*

A2: *It is possible to associate with these two DRSs two distinct syntactic realizations (which are thus linearly ordered); if the syntactic realization associated with the effect DRS precedes that associated with the cause DRS, Explanation applies; with the inverse discursive order, Result applies.*

We are going to show that two opposite discourse theories, RST (Rhetorical Structure Theory, Mann & Thomson 1988) and SDRT in its (Asher & Pustejovsky 2000) version (noted as SDRT-00), lay down Explanation or Result in such ways that A1 is not verified for RST while A2 is not verified for SDRT-00. Next we will propose a solution such that both A1 and A2 are verified.

In RST, discourse relations hold only between two sentences or two clauses (or between two DRSs representing sentences or clauses in the

[12] These axioms are written in SDRT style, however they do not constitute SDRT rules to infer Explanation or Result. More generally, in this section, the issue is to determine which discourse relations should be established in discourses such as (18a) or (18b). The issue is not to compute these discourse relations.

SDRT framework). As a consequence, it is said that Explanation holds between the two sentences of (18a) and Result between the two sentences of (18b). This amounts to Explanation (π_3,π_4) and Result (π_5, π_8) in the DRSs of figures 10-2 and 10-3. However this analysis is inappropriate: the sentence with the causative verb refers *both* to the cause (it indicates who is the agent) and to the effect. Axiom A1 is thus not verified.

Contrarily to RST, in SDRT-00, discourse relations hold between two DRSs without any constraint on their possible syntactic realizations. Therefore, Asher and Pustejovsky can propose to record in the lexical entry of a causative verb that the DRSs for the causing sub-event and the resulting state are linked by Result. This amounts to Result(π_1, π_2) and Result(π_6, π_7) in our examples. However, there is no criterion to choose Result rather than Explanation: the DRSs of the two sub-events involved are only abstract representations without corresponding syntactic realizations. Axiom A2 is thus not verified. The choice of Result rather than Explanation can be explained by the fact that Asher and Pustejovsky chose the discourse relation which follows the chronological order of the events (the cause precedes the effect). However, it is necessary to maintain a distinction between discursive and temporal orders. Recall that the discourses (18a) and (18b), with the same temporal order and different discursive orders, are based either on a particularization or generalization event coreference, which implies assymetrical behavior, e.g., for the insertion of adjuncts.

It is thus not justified to link by Result the DRSs for the causing sub-event and the resulting state of a causative verb. On the other hand, we propose to link these two DRSs by means of the events they represent: we propose a conceptual (non discursive) link, noted as CAUSE, between the events involved, e.g. CAUSE(e_1, e_2) and CAUSE(e_6, e_7) in our examples. This link can be recorded in the lexical entry of a causative verb.

We propose moreover to lay down that the DRSs for the cause sentence and the resulting state are linked by Explanation or Result, e.g. Explanation (π_2, π_4) and Result (π_5, π_7) in our examples. Let us show that both axioms A1 and A2 are verified with this solution; the argumentation will be presented only for Explanation (π_2, π_4), since that for Result (π_5, π_7) is entirely similar.

- For the axiom A1 in Explanation (π_2, π_4), it has to be verified that π_4 represents the cause, and π_2 the effect. This is the case: these DRSs describe the events e_4 and e_2 which are linked by the conceptual link CAUSE thanks to the following implication: CAUSE(e_1, e_2) $\wedge$ $e_1 = e_4$ $\Rightarrow$ CAUSE(e_4, e_2).

  Note that a formal basis for A1 can be given, see A1'.

  A1': *Explanation and Result hold between two DRSs which describe two*
  events linked by CAUSE.

- For the axiom A2 in Explanation (π_2, π_4), it has to be verified that the syntactic realization associated with the effect DRS, i.e., π_2, precedes that associated with the cause DRS, i.e., π_4. This is the case: π_2 has no syntactic realization, however, π_2 is an element of π_3 which does have a syntactic realization, namely the sentence with the causative verb. This sentence precedes the syntactic realization of π_4, the cause sentence. Thereby, by associating with π_2, element of π_3, the syntactic realization of π_3, A2 is verified for Explanation (π_2, π_4).[13]

To sum up, the only way to lay down that (18a) involves Explanation and (18b) Result while having axioms A1 and A2 both verified consists in stating that these discourse relations link the DRSs for the cause sentence and the resulting state. The analyses of (18a) and (18b) which take into account these facts are sketched in figures 10-4 and 10-5.

[13] The same kind of argumentation does not work for Result (π_1, π_2) as proposed by Asher and Pustejovsky, since π_1 and π_2 are both elements of π_3: the only syntactic realizations that can be associated with π_1 and π_2 are not distinct thus not linearly ordered.

Ted cracked the carafe. He hit against the sink	*Ted hit the carafe against the sink. He cracked it.*
$\pi_1=$ e_1Š?ACT(t, c)	$\pi_5=$ e_5Šhit(t, c, s)
$\pi_2=$ e_2Š cracked(c)	$\pi_6=$ e_6Š?ACT(t, c)
$\pi_3= (\pi_1, \pi_2)$ with CAUSE(e_1, e_2)	$\pi_7=$ e_7Šcracked(c)
$\pi_4=$ e_4Š hit(t, c, s)	$\pi_8= (\pi_6, \pi_7)$ with CAUSE(e_6, e_7)
Particularization$(\pi_1, \pi_4) \rightarrow e_1 = e_4$	Generalization(π_5, π_6) with $e_5 = e_6$
Explanation(π_1, π_4)	Result(π_5, π_7)

Figure 10-3. Analysis of (18a)

Figure 10-4. Analysis of (18b)

Summary on causal discourses

We have shown that the causal interpretation of discourses such as (18a) and (18b) is bound to event coreference. Therefore, the discourse relations Particularization and Generalization are also at stake in causal discourses. What is mere Explanation and Result hold neither between the two sentences of the causal discourses nor between the two sub-events of the causative verb. They hold between the DRSs for the cause sentence and the resulting state. The fact that Explanation goes along with Particularization while Result goes along with Generalization allows us to explain why Explanation discourses have a normal behaviour (e.g., any modifier can be inserted) while Result discourses have an awkward behaviour (e.g., heavy constraints on the insertion of modifiers). Explanation and Result are thus not really converse, despite what is generally believed.

4. CONCLUSION

We have studied four types of discourses: on the one hand, PDs and GDs which are close to claboration and reformulation discourses, on the other hand, causal discourses with two linear discursive orders. These discourses are linked by the fact that they all involve an event coreference relation. The type of coreference, particularization or generalization, plays a major role and requires new discourse relations, as recapitulated below.

- (2a) involves Particularization,
- (2b) involves Generalization,
- (3a) involves Particularization and Explanation,

- (3b) involves Generalization and Result.

There are some cases where one can hesitate on the discourse relations involved, depending on subjective factors. To illustrate this point, we give first two examples where there can be no hesitation on the discourse relations involved, before ending with an example whose interpretation depends on subjective factors.

- (25a) is a PD in the style of (16a): *kill* is an hyperonym of *cut the throat*. The temporal adverb adds information not conveyed in the first sentence and its presence is mandatory. (25b) without this adverb sounds deviant (redundant).

  (25) a. Ted cut the throat of the rabbit. He killed it this morning.
  b. # Ted cut the throat of the rabbit. He killed it.

- On the opposite site, (26a) is a result discourse. The event coreference relation at stake has to be of type generalization. It is not the case in (26b) which includes a temporal adjunct in the second sentence and which does not have a causal interpretation (hence the sign #).

  (26) a. Ted hit the rabbit with his wheelbarrow. He killed it.
  b. # Ted hit the rabbit with his wheelbarrow. He killed it this morning.

- Finally, consider (27a) and (27b). Some speakers prefer (27a), while others prefer (27b). The former do no consider the death of the rabbit as inferable from the firing event. In linguistic terms, they interpret (27a) as a result discourse. The latter consider that firing a shot at X ineluctably leads to the death of X. In linguistic terms, they interpret (27b) as a PD.

  (27) a. Ted fired a shot at the rabbit. He killed it.
  b. Ted fired a shot at the rabbit. He killed it this morning.

To put it in a nutshell, the discourse relations involved in a discourse may depend on the speaker's vision of the world.

REFERENCES

Asher, N. (1993), Reference to Abstract Objects in Discourse, Kluwer, Dordrecht.

Asher, N. and Lascarides, A. (1995), Lexical Disambiguation in a Discourse Context, Journal of Semantics **12**(12): 69-108.

Asher, N. and Lascarides, A. (2003), Logics of Conversation, Cambridge University Press, (to appear).

Asher, N. and Pustejovsky, J. (2000), The Metaphysics of Words in Context, Journal of Logic, Language and Information, (submitted).

Corblin, F. (1994), La condition de nouveauté comme défaut, Faits de langues (Presses Universitaires de France), (4): 147-160.

Corblin, F. (1995), Les formes de reprise dans le discours, Anaphores et chaînes de référence, Presses Universitaires de Rennes, Rennes.

Danlos, L. (2000), Event Coreference in Causal Discourses, In P. Bouillon and F. Busa (eds.) Meaning of Word, Cambridge University Press.

Danlos, L. (2001), Event Coreference Between Two Sentences. In H. Bunt (ed.) Computing Meaning (volume 2). Kluwer, Amsterdam.

Fodor, J. (1970), Three reasons for Not Deriving kill from cause to die, Linguistic Inquiry.

Hobbs, J. (1979), Coherence and Coreference, Cognitive Science, (3): 67-90.

Humphreys, K. and Gaizauskas, R. and Azzam, S. (1997), Event coreference for information extraction. In Proceedings of the Workshop on Operational Factors in Pratical, Robust Anaphora Resolution for Unrestricted Texts, 35th Meeting of ACL, Madrid.

Kamp, H. and Reyle, U. (1993), From Discourse to Logic, Kluwer Academic Publishers, Dordrecht.

Mann, W. and Thomson, S. (1988), Rhetorical Structure Theory: Toward a functional theory of text organization. Text: An Interdisciplinary Journal for the Study of Text **8**(2).

Moens, M. and Steedman, M. (1988), Temporal Ontology and Temporal Reference, Computational Linguistics **14**: 15-28.

Pustejovsky, J. (1991), The syntax of event structure. In Lexical and Conceptual Semantics. Elsevier Science Publishers, Amsterdam.

Pustejovsky, J. (1995), The generative Lexicon. The MIT Press.

Rossari, C. and Jayez, J. (1996), Donc et les consécutifs ; des systèmes de contraintes référentielles. Lingvisticae Investagiones **20**(1): 117-143.

Roussarie, L. (2000), Un modèle théorique d'inférences de structures sémantiques et discursives dans le cadre de la génération automatique de textes, Université Denis Diderot, Paris 7, Thèse de doctorat en linguistique.

Schank, R. (1975), Conceptual Information Processing, Amsterdam: North Holland.

Spejewski, B. and Carlson, G. (1991), Reference Time Relations. In Proccedings from the Eastern States Conference on Linguistics, Baltimore.

Webber, B. (1988), Discourse deixis: Reference to discourse segments. In Proceedings of ACL'1988, pp. 113-123

Chapter 11

RATIONALITY IN CONTEXT

J. Francisco Morales
UNED, Spanish National University for Distance Teaching

1. INTRODUCTION

The "rational-irrational" dichotomy (Jackson, 1988) has deep and far-reaching consequences. Rational behavior, insofar as it is adjusted to the rule of rationality, is regulated, structured, and predictable. Any departure from rationality, especially if it is serious, should be expected to drive behavior towards the opposite end of lack of regulation, structure, and predictability.

The dichotomy underlies Asch's (1948) well-known research on the evaluation shifts of a famous text as a function of their alleged authorship. In the early influence studies it was found that samples of American College students tended to agree with this text when it was attributed to Jefferson and to disagree with it when told that Lenin was its author. To explain why the same content was differently evaluated, subjects' irrationality was generally postulated. It was speculated that the subjects, instead of focusing on the text content, were influenced (irrationally) by the author's prestige (of course, Lenin's prestige was not very high in American Colleges). Asch conducted his own research to demonstrate that irrationality was out of question. His subjects, also American College students, endorsed more a text when attributed to Jefferson than to Lenin, but this was due, as the research clearly showed, to a "cognitive reorganization" of the meaning of the text itself, not to an "uncritical evaluation" (see Jackson, 1988, p. 31).

The rational-irrational dichotomy is still with us. Kinder (1998) has called recently attention to the so-called "Rational Public Opinion" issue.

K. Korta and J. M. Larrazabal (eds.), Truth, Rationality, Cognition, and Music, 211-221.

Systematic surveys on President's popularity in the US repeatedly show dramatic ups and downs. How could one explain these profound changes of the "President's standing in the public"? It is tempting, as some have done, to appeal to the public's plain irrationality. However, Page and Shapiro (1992), much like Asch, have shown that this is not the case. Through a careful and detailed scrutiny of many surveys and other kinds of data, they reached the conclusion that the public reacts to the political situations in "sensible, reasonable ways", in other words, that it is "rational", and behaves accordingly (see Kinder, 1998, p. 798).

2. VIEWS OF RATIONALITY

But, why rationality should be associated with structure and predictability of behavior? Or, going a step further, which is the exact meaning of "rationality"? Several answers have been advanced.

For Abelson and Levi (1985, p. 232), rationality has to do mainly with decision-making. Synonymous terms would be "optimality" or "functionality". Correspondingly, "irrationality" would be equated with error and maladaptiveness. For these authors the issue of rationality is closely related to "purposive human behavior", implying that such behavior is usually "goal-directed" and that there are better (less time-consuming, less error-prone) and worse (more time-consuming, more error-prone) ways of getting the goal.

Two consequences follow from Abelson and Levi's perspective: a) to inquire about rationality of a given human behavior, a contrast should be established between the performance or behavior actually observed and the performance or behavior which could be considered optimal for attaining the goal; b) social psychology should restrict the rationality question to decision-making, and refrain from extending it to other areas of research and theory such as "affiliation, aggression, conformity" and the like.

For some social psychologists, Abelson and Levi's position is unduly restrictive in limiting rational choice to decision-making, which is, obviously, a small segment of social behavior. Snow and Oliver (1995, pp. 583-584), following a different path, widen the scope of rational choice, arguing that it focuses on "the role of different cognitions in determining behavior or the mechanisms linking cognitions and behavior" (1995, p. 583). From their perspective, theoretical developments such as control theory, learning theory and relative deprivation theory are "variants of rational choice". It is granted that cognitions precede behaviors and that choices are conscious, intentional, and rational. To this, two assumptions are added: a) that people are expected to strive for benefits and to minimize costs, and b) that there is cognitive processing of information on the likelihood of benefits and costs of alternative courses of action followed by a conscious choice of one of these courses.

Along similar lines, Pfeffer (1998, p. 743) speaks of the "prominence" of economic models in social sciences, equating "economic" with (presumed) rationality of behavior or maximization of utility or preferences. Social relationships, even organizations and societies, are assumed to pursue and attain efficiency, in order to avoid failure and eventual disappearance.

Alan Fiske and colls. (1998) link rational choice with Market Pricing (MP), one of the four ways elaborated by the first author to generate and understand relationships. MP means "sense of proportion", calculation of "social values as ratios", is exemplified by prices, rents, interests, and utilitarian moral calculations, among others, and is the typical relationship in which vendors, customers, and stockbrokers are involved. It entails "cost-benefit" analysis. Its presence in Western, capitalist countries is so ubiquitous that it invades other areas of human sociability. It is not confined to economic areas. In Fiske's terms, in the mentioned countries, "people evaluate the costs and benefits of the outcome possibilities for each action they might take and weight these by their probability of occurrence. The action with the highest probability-weighted ratio of costs to benefits is the one chosen" (1998, p. 952).

3. ARE THERE LIMITS TO RATIONALITY?

Abelson and Levi (1985, pp. 233-234), acknowledging that rationality has its limitations, proceed to establish a distinction among several types of rationality. There is, in the first place, "the corrigible rationalist", characterized by the tendency to ignore the criterion of "optimum task performance", for example, the person who does not follow logical rules when trying to arrive at a judgment; the "bounded rationalist", who prefers to satisfice than to optimize, that is, the person who is contented with less than the maximum he or she could get, the "error-prone intuitive scientist", who tries to be rational but ends up short of this goal due to undetected subtleties of the situation, in other words, his or her fallibility becomes evident when the complexity of the situation increases, and the "slave to motivational forces" who is driven by emotionality or passion and loses control.

Fiske and colls. (1998, p. 953) think that the rationality becomes limited when social behavior explanations are accepted only if formulated in exclusively or predominantly economic terms. True, there are certain cases where economic predictions hold, but, in general, they only work for people who have been trained specifically to follow economic principles. The authors quote Kahneman and Tversky in support of their position: people, in their choices, do not follow "the fundamental axioms of cost-benefit theory." Here are some of the major departures from the theory: a) outcomes are valued on a relative scale, not on an absolute one, b) usually "opportunity costs" are not recognized —that is, people tend to ignore that they are losing something when making a particular choice, c) they are unaware that they are victims of a "sunk cost trap", that is, if they pay for any activity, they are committed to it even after realizing that it has lost any value for them.

Tetlock (1998, p. 886) has found limitations of rationality in his studies on political judgment and decision-making. Somewhat paradoxically, accountability tends to be associated with erroneous decisions. For instance, on certain occasions, a politician may be accountable to a deeply disoriented and/or ignorant constituency, or he or she is under strong pressure from an ingroup (the case of groupthink), and, more generally, there are the "perils of the foolish audiences", those, that in Tetlock's terms, are "fickle, superficial, or impulsive", and push the politician to take faulty decisions.

According to Fiske and colls. (1998, p. 953), people find sometimes difficult to behave "rationally", even when trying to do so. They indicate that "psychological research shows that even Americans and other Westerners think in this MP, economic mode only in certain circumstances, with many limitations, and as a function of their education". Snow and Oliver (1995, pp. 585-586) present another, very well known, example of this difficulty: people participating in a collective movements tend to "attribute themselves as individuals" the efficiency of the whole movement. It is a clear case of over-attribution to the self, which the authors call "transformation of efficacy".

4. EXTENDING RATIONALITY

An attempt to "extend" rational choice theory, referred to by Snow and Oliver (1995, pp. 584), has to do with the notion of incentives. While for traditional rational choice only "material" ones count (e.g. "salaries, insurance programs and threats of physical or economic retaliation"), other kinds are currently being considered. Solidary incentives have to do with relational aspects. They may be positive, like "praise, respect, and friendship", or negative, like "shame, contempt and ostracism". Purposive incentives refer to internalized norms and values, for example, doing what one thinks is right. As it will be shown later in this paper, solidary and purposive incentives link rational choice theory to other areas of interest for the social sciences, for instance, culture and socialization.

Yamagishi (1995, p. 313) presents another extension attempt based on a reconceptualizacion of "social dilemmas". The mere fact of organizing a social movement is a social dilemma in itself. Consider the case of a given community, involved in a collective effort to stop the building of a dam. If the community does not succeed in mobilizing a sufficient number of its members, and, as a result, the dam is finally built, everybody in the area will suffer, but especially those who have participated in the mobilizations, because, in addition to the environmental damage caused to the area where they live, they have wasted their own resources, time and energy. But what if the sufficient number of people have participated and succeeded in stopping the building of the dam? Even those who have not participated will benefit. Under these conditions, not joining the movement will be the dominant choice. However, this dominant choice means environmental damage caused to the area where they live (here lies the "dilemma").

Perhaps the fundamental question involved in a social dilemma (Yamagishi, 1995, pp. 317-318) is what moves people to act cooperatively (instead of competitively). For some authors, people will act cooperatively only if they care for others. These researchers follow a "social motivation-social orientation" line and focus on individual differences in altruism, since they assume that people high in altruism are more ready to cooperate in a social dilemma. It is "rational" for them to do so: others' well-being is more important for them. Other researchers focus instead on the existent structure of incentives. They assume that through the modification of such structure, for example, by providing rewards for cooperation and imposing fines for non cooperation, it will become possible to increase the level of cooperative behavior.

In a more positive vein, still other researches believe that cooperation is possible when one of these three conditions are met: a) people realize that their self-own interest is contingent upon mutual cooperation and trust each other (goal expectation approach), or b) they realize that the costs incurred in inducing others' cooperation are inferior to the benefits brought by this cooperation (strategy approach), or c) they force cooperation through the use of sanctions (norm-formation approach).

5. RATIONAL CHOICE AND CULTURE

Is it possible to reconcile rational choice theory with the influence of culture? We think it is and will try to explain how, using an example of a well known cross-cultural study on conflict management strategies by Ohbuchi and Takahashi (1994), who, following Falbo and Peplau (1980), present a two dimensional model of power strategies resulting in four main categories:

a) Bilateral indirect, for example, suggesting, ingratiation, impression management and appeasing,
b) Bilateral direct, for example, persuasion, bargaining and compromise,
c) Unilateral indirect, for example, sadness, crying, detour and deception,
d) Unilateral direct, for example, asking, supplicating, assertion, threat, coercion and criticism.

Since Japanese and American subjects participated in the study, Ohbuchi and Takahashi (1994), added a fifth strategy: avoidance. Participants were asked to evaluate the "perceived effectiveness" of each kind of strategy, from 1 (not at all effective) to 5 (perfectly effective). No differences in the evaluation of "perceived effectiveness" were found between Japanese and American subjects. Both perceived direct bilateral strategies as the most, and avoidance as the least, effective.

Differences between Japanese and Americans appeared in preferences, the latter (Americans) showing significantly higher preferences for direct bilateral strategies. Differences showed up also in the percentage of engaged strategies, Japanese engaging in higher avoidance and indirect bilateral strategies (and lower direct bilateral and indirect unilateral strategies) than Americans.

With respect to the differences between desired and engaged strategies, it was found that they were significant in both groups. Specifically, direct bilateral strategies were significantly less enacted than desired while avoidance was significantly more enacted than desired. At the same time, the differences (desired vs engaged strategies) were larger among Japanese.

Ohbuchi and Takahashi (1994) felt the need to introduce a further distinction between overt and covert conflicts. A conflict is overt when an offending party is perceived as being aware of interfering with the other, while it is covert when the offending party is not aware, and it depends on the victim to make the conflict overt or not. But what could be the motives of the victim to make a conflict covert? In principle, there are several possible motives:

a) Maintenance of relationships,
b) Prevention of conflict escalation,
c) No solution (no idea how to resolve the conflict),
d) Sharing responsibility (one considers him/herself responsible of the situation),
e) Nonseriousnes of the conflict,
f) Others.

Participants in the study reported on the whole 238 covert conflicts, 79% of which corresponded to Japanese. In fact, 65.5% of the total amount of conflicts reported by Japanese were covert, the corresponding figure being a mere 26.7% for Americans.

Unlike Americans, Japanese reported two main reasons for making conflicts covert: maintenance of relationships, in the first place, and feelings of shared responsibility, in the second. The first reason reflects Japanese fear of disturbing social harmony or their perception of social pressure toward maintaining it. The second reason reflects Japanese interdependent selves: even being the victims in a conflict, they will feel that they are disturbing the social harmony and, therefore, will consider themselves partly responsible for the conflict.

It should be remembered that Japanese did not differ from Americans in the evaluation of perceived effectiveness of the strategies. This precludes an interpretation based on the idea that, for the Japanese, the annoyance caused by covert conflicts was tolerated because of the expectation of an eventual benefit. If this were the case, they would have evaluated avoidance more positively in terms of effectiveness, which they did not.

The alternative interpretation would contend that Japanese are apt to overreact emotionally to other people's uneasiness. This is in line with findings from cross-cultural social psychology, according to which Japanese share a belief that a request, no matter how reasonable or polite, implies discontentment or criticism, damages the other's pride or social face, and implies a danger for the future of actor's relationship with him or her. Consequently, for the Japanese, it may be very difficult to overcome such a fear and engage in active resolution strategies. So, avoidance is a rational strategy, after all, but with a rationality that could be called "cultural", and may subsumed under the "solidary" incentives proposed by Snow and Oliver (1995).

6. RATIONAL CHOICE AND RELATIONAL CONCERNS

The obedience to authorities, a question which has been thoroughly analysed by Tyler and Lind (1992), is usually explained in terms of public choice theory —and social exchange theory (SET): people are assumed to obey authorities simply as the best way to maximize their outcomes. This is a direct application of the self-interest principle.

According to instrumental or social exchange theories, procedural justice is determined largely by outcomes, concerns, and interests (it is worth remarking that these outcomes, concerns, and interests are external both to the procedure and to the interaction with the authority). Tyler and Lind (1992) argue, on the contrary, that procedural justice is best explained on the basis of "relational theory", in other words, that it is rather a function of the perceptions of the relationship between authority and those subject to his or her decision. They contend that procedures are evaluated in terms of their implications for group values and for how the person is viewed by the group using the procedures, while reactions to procedures have to do with long-term social relationship to authorities.

Tyler and colls. (1998) provide empirical evidence in support of their argument. They hypothesize that people will attend more to relational concerns in in-group interactions than in out-group interactions, since they will be likely to be more concerned with relational aspects when sharing some salient social category with the person involved in the conflict.

In their study there were four ethnic categories of participants: 45 African American, 56 European American (Whites), 117 Asian American, and 58 Hispanic Latino. All of them were employees of a State University, representing all types and levels of jobs (only members of ethnically based work associations were selected for the study). Their task was to provide a description of a recent conflict with his or her work supervisor, i.e., a situation in which they had spoken to their supervisor about any of a number of issues (such as decisions about how work should be done, discussion of pay, promotion, resolving a dispute with a customer, coworker, and other issues). They had to answer two questionnaires. The first one, or instrumental, included questions such as how was the outcome?, how does outcome compare to expected outcome?, how much better or worse off were you compared to previous situation?, how does outcome compare to outcomes received by others in similar situation?, and in terms of your outcome, did you win or lose? The second, or relational, included the following questions: how dignified the treatment?, how favourable, how polite the treatment?, how much concern for your rights?, how openly issues were treated?, how honest the supervisor?, how much of an opportunity to describe the problem?, how seriously were taken your needs?

Conflict was classified depending on the match of ethnicity. For example, Hispanic supervisor – African American worker is a mismatch, while Hispanic supervisor – Hispanic worker is a match. On the whole, there were 195 mismatches and 85 matches.

The dependent or criterion variable in the regression equation was the decision acceptance, i.e, was the worker willing to voluntarily accept the decision of the supervisor? There were several independent (or predictor) variables: instrumental judgments about the experience, relational judgments about the experience, interaction of instrumental judgments x within/between group distinction, interaction of relational judgments. x within/between group distinctions. In order to obtain support for the hypothesis and relational model, relational judgments must show stronger regression coefficients in the ethnic-match condition. The results of simultaneous regressions show this to be the case.

Tyler and colls. (1998) conclude from these findings that people's orientations to authorities depend above all on the treatment they get, the effect being stronger when identity issues are salient. Instrumental concerns are less important.

6.1 The "myth of self-interest"

Tyler and colls. (1999) borrow the label "myth of self interest" from a paper by Miller and Ratner (1998). It is generally assumed that self-interest beliefs consist of two different beliefs:

a) that everybody, including oneself, shapes his or her behaviors so as to maximize individual self-interest,
b) that self-interest is always defined in terms of material gains and losses and exchange of material goods.

Research on reaction to realworld experiences (e.g. police or courts) seems to show that the prevalence of self-interest is overestimated. People are more concerned about issues of morality and procedural fairness than about best material outcomes. It is true that studies of a priori preferences for one conflict resolution procedure over another have found strongest support for self-interest antecedents. However, when the focus is on post hoc evaluation of experiences, relational antecedents have been found to be more decisive. In this respect, Tyler and colls. (1999) distinguish between the previous beliefs and the actual experience with a given procedure. Unlike beliefs, the actual experience is likely to engage a set of social identity processes associated with relational concerns.

Something similar happens in studies on mediation. Conflicting parties are typically reluctant to choose mediation, since they think that mediation lessens third party control as well as makes problematic the attainment of any given outcome. But mediation, at the same time, engenders feelings of social engagement and involvement. Indeed, after having been experienced, mediation is usually evaluated quite positively by disputing parties and produces decisions likely to be satisfactory and accepted.

So it seems that people's preferences are shaped by different psychological dynamics than their evaluations. In the pre-experience calculus, social considerations are underconsidered. People do not recognize that a key input into their post-experience satisfaction is the quality of treatment they receive from others.

7. CONCLUSION

People act rationally in most circumstances, even when their behaviors may seem irrational at first sight. In general, research areas reviewed in this paper suggest that the fundamental rationality of behavior shows up when the context is borne in mind. Take Ohbuchi and Takahashi's research (1994) for an example: Japanese subjects chose a conflict management strategy they themselves did not consider efficient. Was this irrational? On the contrary. What they did was to follow the strategy most favored by Japanese culture.

The context, as defined by the person, is an essential ingredient of behavior. Paying attention to it is a necessary step to move beyond the rational-irrational dichotomy to a more fruitful, and adequate, conceptualization of human behavior functionality.

REFERENCES

Abelson, R.P. y Levi, A., 1985, Decision Making and Decision Theory, in G. Lindzey and E. Aronson (eds.), The Handbook of Social Psychology, New York, Random House, Third Ed., Vol. I, pp. 231-309

Asch, S.E., 1948, The doctrine of suggestion, prestige and imitation in social psychology, Psychological Review, 55, pp. 250-277

Falbo, T. and Peplau, L.A., 1980, Power strategies in intimate relationships, Journal of Personality and Social Psychology, 38, pp. 618-628

Fiske, A.P., Kitayama, S., Markus, H.R. and Nisbett, R.E., 1998, The cultural matrix of Social Psychology, in D. T. Gilbert, S.T. Fiske and G. Lindzey (eds.), The Handbook of Social Psychology, New York, McGraw-Hill, Fourth Ed., Vol. II, pp. 915-981

Jackson, J.M., 1988, Social Psychology, Past and Present: An integrative approach, Hillsdale, L. Erlbaum

Kinder, D.R., 1998, Opinion and Action in the Realm of Politics, in D. T. Gilbert, S.T. Fiske and G. Lindzey (eds.), The Handbook of Social Psychology, New York, McGraw-Hill, Fourth Ed., Vol. II, pp. 778-867

Miller, D.T. and Ratner, R.K., 1998, The disparity between the actual and the assumed power of self-interest, Journal of Personality and Social Psychology, 74, pp. 53-62

Ohbuchi, K.I., and Takahashi, Y., 1994, Cultural Styles and Conflict Management in Japanese and Americans: Passivity, Covertness, and Effectiveness of Strategies, Journal of Applied Social Psychology, 24, 15, pp. 1345-1366

Page, B.I. and Shapiro, R.Y., 1992, The Rational Public, Chicago, University of Chicago Press

Pfeffer, J., 1998, Understanding Organizations: Concepts and Controversies, in
D. T. Gilbert, S.T. Fiske and G. Lindzey (eds.), The Handbook of Social Psychology, New York, McGraw-Hill, Fourht Ed., Vol. II, pp. 733-777

Snow, D.A. and Oliver, P.E., 1995, Social Movements and Collective Behavior: Social Psychological Dimensions and Considerations, in K.S. Cook, G.A. Fine and J.S. House (eds.), Sociological Perspectives on Social Psychology, Needham Heights, Allyn and Bacon, pp. 571-599

Tetlock, P.E., 1998, Social Psychology and World Politics, in G. Lindzey and E. Aronson (eds.), The Handbook of Social Psychology, New York, Random House, Fourth Ed., Vol. II, pp. 868-914

Tyler, T.R., Huo, Y.J. and Lind, E.A., 1999, The two Psychologies of Conflict Resolution: Differing Antecedents of Pre-Experience Choices and Post-Experience Evaluations, Group Processes and Intergroup Relations, 2, 2, pp. 99-118

Tyler, T.R. and Lind, E.A., 1992, A relational model of Authority in Groups, in M.P. Zanna (ed.), Advances in Experimental Social Psychology, San Diego, Academic, Vol. 25, pp. 115-191

Tyler, T., Lind, E.A., Ohbuchi, K.I., Sugawara, I. and Huo, Y.J., 1998, Conflict with Outsiders: Disputing within and across cultural boundaries, Personality and Social Psychology Bulletin, 24, 2, pp. 137-146

Yamagishi, T., 1995, Social Dilemmas, in K.S. Cook, G.A. Fine and J.S. House (eds.), Sociological Perspectives on Social Psychology, Needham Heights, Allyn and Bacon, pp. 311-335

Chapter 12

INDIVIDUAL AND COLLECTIVE RATIONALITY IN A SOCIAL FRAMEWORK

Luis A. Pérez Miranda
ILCLI, The University of the Basque Country

Abstract: Our aim is to analyse and discuss some of the main *criteria of rationality* as they have been proposed by rational choice theory and game theory. The paper is divided into two parts. Part one is devoted to problems of individual decision-making from the viewpoint of instrumental rationality, and part two focuses on decision-making problems that arise when we try to extend rational choice theory from individual decisions to strategic decisions with two or more individuals. Strategic decisions bring us to game theory. Generally speaking the aim of this paper is to provide an outline of some of the contributions and limitations of instrumental rationality in facing decision-making problems of special interest for social and cognitive sciences.

INTRODUCTION

First, we'll briefly introduce the very basics of standard rational choice theory also known as decision theory. No attempt will be made here to summarise more than a few of decision theory's main contributions to the problem of rational choice. Decision theory is basically a normative theory that tells agents how they should choose among different alternatives if they are to be rational. From this normative perspective rational choice is seen as that which selects the most effective means to satisfy the agent's preferences, where those preferences are consistent and complete. This theory has also set out to describe an agent's relevant attitudes using appropriate probability and utility functions. However, experimental results

K. Korta and J. M. Larrazabal (eds.), Truth, Rationality, Cognition, and Music, 223-241.

indicating that people systematically violate the axioms of utility theory have led either to substantive modification or to abandoning it altogether because of its explanatory inadequacy. Moreover, leaving aside the philosophical or theoretical discussion around rational choice theory, we cannot forget that there are also informational, architectural, physical, and resource limitations that constrain notably its use in the modelling of practical rationality. *Second*, we'll show that for cases of interactive decision-making the use of some sort of *strategy* becomes necessary. This is the realm of game theory. For the case of decision-making problems that involve more than one agent, game theorists have developed a very sophisticated analysis of "equilibrium" strategies in games, but precisely why it is rational for players to choose these strategies remains a matter of dispute. We'll present two non-cooperative games that represent relevant situations of social interaction and have deserved the attention of social scientists. In particular we'll pay attention to two basic games: *coordination* and the *prisoner's dilemma* games. In this part of the paper we want to emphasise that non-cooperative game theory focuses on how cooperation can emerge as rational behaviour from situations of non-binding agreement. *Third*, a few words will be said about the inconvenience of using the original rational choice paradigm to account for collective (or social) choice. Our view is that rational choice theory has some important implications in social domains, but it still seems inadequate in explaining relevant aspects of social action.

1. INDIVIDUAL DECISION-MAKING IN A SOCIAL FRAMEWORK

1.1 The Basic Model of Rational Choice Theory

The apparatus of the rational choice theory applies to a wide variety of theoretical and practical issues: rational strategic interaction, game theory, social choice and welfare economics, the theory of microeconomic phenomena, the building up of cognitive architectures, etc. The decision-theoretical model stems from a refined theory of utilitarianism. In this model it is assumed that an agent with a variety of goals is capable of comparing the satisfaction of these goals so as to come to an overall assessment. Rational choice theory assumes that preferences (or desires) can be ordered on a single scale by comparing the "pleasures" of satisfying them. The name given to this measure is "utility". Therefore the instrumentally rational agent can compare the utility generated by each action and act so as to maximize his utility.[1]

In a decision-theoretical framework deliberation can be understood as the evaluation of available lines of action in terms of their outcomes, which may depend on circumstances the agent can neither predict nor control (Jeffrey, 1983). That is, the agent's notions of the probabilities of the relevant circumstances together with the utilities of the possible outcomes are represented numerically in order to determine the expected utility value for each of the alternatives under consideration. Von Neumann and Morgenstern's theory applies to agents who must choose among a set of feasible actions without being able to predict with certainty what the outcome of a given choice will be. According to their theory if individual preferences satisfy a given set of axioms, they can be represented by the expectation of some utility function.[2] The numerical probabilities of the circumstances and utilities of the outcomes would reflect the agent's actual beliefs and preferences (von Neumann & Morgenstern, 1944; DiFinetti, 1975; Savage, 1972; Skyrms, 1980; Jeffrey, 1983; Hargreaves *et al.*, 1992). The ideal decision-maker would know all behaviour alternatives and the exact outcome of each alternative; this decision-maker would also have a complete preference order for those outcomes. More specifically, in this framework a decision problem arises when we have a finite set A of alternative acts among which we must decide, a finite set Γ of exhaustive and exclusive possible outcomes, a function U assigning numerical utilities to the possible outcomes, and we have reasonable beliefs about the values of the Prob (O/A) of each possible outcome on each act. In other words, we have to choose the act with the highest expected utility.

The Expected Utility Value of an act is defined as follows:

$$EUV(A) = \sum_{O \in \Gamma} U(O) \cdot Prob(O/A)$$

Thus, following the *rational choice principle* to be rational is to choose an alternative of maximum expected utility.

[1] Modern approaches in social sciences directly assume that individuals have preferred alternatives and that the integration of these preferred alternatives is revealed in a *preference ordering* which determines action. According to this approach rational agents choose maximally preferred alternatives.

[2] The set of axioms concerning individuals' preferences over lotteries are: (1) Reflexivity; (2) Transitivity; (3) Completeness; (4) Continuity; (5) Reduction; and (6) Monotony. When conditions (1)-(6) hold, that ordering can be represented by a utility function which is unique up to a strictly increasing transformation. In terms of the rational choice principle, an agent will be rational if he chooses maximally preferred alternatives according to the aforementioned conditions.

1.2 Scope and Limits of the Decision Theoretical Model

It could be argued that it is useful for agents that have to deal with hard practical problems to think of them in decision-theoretical terms. But is that method good advice? Do agents make better decisions using such a method or not? As we shall see, it is not obvious that the suggested method is always good advice. There are many aspects, such as those related to the capacities of agents, the complexity of reasoning procedures or the violation of some of its utility axioms, which undermine the application of rational choice theory as a full theory of practical rationality.

Limited agents and knowledge incompleteness

On the one hand, in everyday life the information available at the moment of dealing with a problem is not usually rich enough to enable the agent an optimal decision in terms of expected utility. A decision-maker with bounded rationality would not know all the alternatives, would not know the exact outcome of each, and would not have a complete preference order for those outcomes. Human beings as opposed to ideal reasoners do not usually have clear or reliable information about probabilities of the different outcomes.[3]

On the other hand, rationality in the ideal theory considers only whether the results of choices best satisfies agents' preferences, while rationality in limited agents also considers whether the agent makes good choices in the process of deciding how to apply its efforts in reasoning toward a decision (cf. Doyle, 1992). Acting in order to obtain useful information is an important part of rationality. But this does not mean, however, that we can extend our search for information indefinitely. Sometimes the cost of processing information about alternatives outweighs the rewards of an optimal choice; under several circumstances it is better to make a choice on time, and get on with adjustments afterwards if things do not work out at a satisfactory level. Most realistic tasks have a deadline that cannot always be postponed. So, we have to accept that knowledge incompleteness is part of the human condition, and many of our decisions must somehow rest on beliefs that at most we may justify but not logically deduce.

Coherentism

Some problems with this model are not related to the capacities of agents, but to the properties of the rational choice paradigm itself. Why should we

[3] Instead, this situation doesn't preclude us from going on with our processes of practical reasoning and drawing out conclusions in a rational way given our beliefs and desires: Inductive reasoning and defeasible reasoning are valid examples of it.

expect that any simple quantitative model could be used as a reasonable description of an agent's behaviour? "The fundamental results of decision theory directly address this question, by showing that any decision-maker who satisfies certain intuitive axioms should always behave so as to maximize the mathematical expected value of some utility function, with respect to some probability distribution" (Myerson, 1991, p. 5). For classical instrumentalism rationality of actions remains subordinated to agent's intrinsic desires and it is only governed by coherence criteria. The coherence or consistency of agent preferences may certainly be a desirable property, but one thing is to be coherent according to certain formal conditions, and another thing is to be rational. Something other than coherentism is probably needed to understand how and why an agent pursues the set of goals that he does. This last point takes us to the question of whether the notion captured by the rational choice theory fits accurately or not our intuitive notion of idealised rationality. The answer seems to be conditioned by our own intuitions and, because of that, it is the subject of an open debate. Why should we use rational choice theory as a tool for understanding rationality? Is the rational choice principle the right way of characterising ideal rationality? Is this principle the primary motor of deciding systems?

Minimal rationality

Rationality of course cannot require an agent to think of all of the most efficient and adequate means of reaching goals. Even if a course of action could be defeated, according to the agent's set of beliefs and motivations, it doesn't always imply that the action is necessarily irrational. Rationality just requires an agent engaging in a minimally satisfactory plan search that will lead to the achievement of desired goals. Often, for example, there will be a more efficient procedure for doing something unknown to a number of agents, though used by more experienced ones. But if little is lost, the procedure used for the occasion might still be considered minimally rational (cf. Cherniak, 1986). In the building up of cognitive architectures, then, we don't formulate the objective of rationality as one of "maximizing" the achievement of agents' adopted goals by rendering situations according to those goals. In fact what we take as a normative pattern depends to a large extent on what we consider to be possible for artificial or natural agents given their abilities, potentialities, and interests. Moreover deep deliberation is not always available on certain occasions, and other selecting mechanisms (heuristics) must be employed in deciding to adopt a goal. It is common for agents to decide to adopt a goal conditionally and without further deliberation when they realise that a certain condition is fulfilled. For instance, condition-action rules, which are necessary owing to agents' resource limitations, drive an important amount of agents' actions.

Utilities, probabilities, and preferences

The sets of outcomes over which people make decisions are not always strictly comparable with utilities insofar as they are not measured, say, in an economic fashion. Utilities and probabilities are different things. For instance, we can ask about the kind of units in which the result of calculating the expected utility value of an act is measured. Moreover, it is not easy to think of a single scale on which to accommodate the utilities and probabilities linked to outcomes in different areas (cf. Collins, 1994). The existence of a utility function means that all values can be combined into a single value system of desirability, but this may not always be possible.

There are situations where the agent's choice may depend on the set of options and/or the prevailing norms. This means that preferences can change for reasons exogenous either to the own attributes of the goal itself or the rationality of agents. Such situations with context dependent preferences create a significant problem for those approaches to the axiomatic interpretation (of the rational choice under uncertainty) that require the primitives of choices are defined independently of the choice set (cf. Hargreaves, 1998). For all these reasons it is not surprising that people's preferences are not consistent with the decision theory axioms concerning utilities and probabilities. Thus, decision theory "is not at first glance, an adequate descriptive theory of the preferences people have or the way in which they make decisions" (Kyburg, 1991, p. 133).

Lack of experimental adequacy

Some of the most interesting recent research on decision theory has been motivated by empirical results that reveal systematic violations of expected utility axioms. The paradoxes of Allais (1953) and Ellsberg (1961) showed how the von Neumann-Morgenstern independence axiom is sometimes violated. Tversky & Kahneman (1974) also described certain problems through experiments where decision-theoretic principles (axioms of the theory) failed to obtain because they were not applied by the agents in the process of resolving common decision-making situations. Dissatisfaction results with models that adhere to the "perfect rational man" have been the basic motivation for the proposal of new models. For instance, nowadays there are models of rational choice that drop the property of transitivity on preferences without necessarily rendering agents irrational. "Insisting on transitivity would rule out many of the new theories of utility maximization that decision theorists have shown to exist" (Anand, 1993, p. 555). Although attempts to replace the basic model of the rational choice paradigm with alternative decision models are not new (just remember the influence of papers written by H. Simon in the mid-1950s), it should be noticed that only recently have these attempts affected mainstream the corpus of the social sciences. Generally speaking, formal models of bounded and situated rationality might be considered accurate if they were able to capture the evidence offered by experimental observations about human behaviour. This last point of view has recently been emphasized by Rubinstein (1998, p. 193):

> "Overall, I agree with Herbert Simon that the departures from the rational man paradigm have to be based on some empirical (or experimental) observations about the basic motives that drive decision makers. The number of deviations from a fully "rational" model is so large that we must content ourselves with the study of extensions that make more sense. This underlies the relevance of some of the cognitive psychological literature, especially work that explicates the clear and simple motives that often appear in real decision making".

Goals and dynamic contexts

As we have already seen for instrumentalism something is rational relative to given goals, ends, desires, and utilities when it is causally effective in satisfying them. "But the notion of instrumental rationality gives us no way to evaluate the rationality of these goals, ends, and, desires themselves, except as instrumentally effective in achieving further goals taken as given" (Nozick, 1993, p. 139). To be sure, rational choice theory overlooks the reasons or motivations for which goals are proposed, say, by an options analyser. That is perhaps one of the reasons why classical instrumentalism cannot explain agency in autonomous agents that behave in dynamic contexts. In such contexts even if the system ought to aim for goal stability, options need to be frequently challenged in a high degree. For instance, when computing the expected utility value of a potential goal, searching or planning are often required, because an agent does not know what the value of a goal is until he knows where he can get to once that goal has been attained, i.e., the derived scenario. As planning systems, agents cannot ignore the essential role played by goals in explaining why and how actions are planned. We think that goal assessment is necessary for agents to be able to select just those goals which it would be "rational" to pursue given their overall interests.[4] But in rational choice theory agents are not

[4] Means-ends reasoning is the process by which an agent chooses a particular goal or end that needs to be satisfied, selects the plan or means of achieving it, and finally intends the plan action. This plan action then leads to the adoption of further sub-goals and further sub-plans. This process ultimately stops when the agent has no more goals to satisfy. The process of choosing relevant goals among a set of alternative conflicting goals is more complex than it might seem at first. "Social animals or machines need goal generators that

considered planning systems, and their actions are explained directly in terms of the rational choice principle. Rational choice theory is concerned with finding answers to questions like: What does it mean to choose rationally? How should we make choices when the consequences of our actions are uncertain? But decision-makers also make decisions about how and when to decide; so we have to look for models that will take account of such decisions as well (cf. A. Rubinstein, 1998). A rational agent's choice among potential goals does not just depend on their mere desirability, but also on how they should be obtained and on the likely consequences of goal satisfaction and plan execution.

Sometimes agents are faced by a conflict among reasons for adopting goals. For instance, the agent might have to choose between a goal he desires and another he ought to carry out because it is morally valuable. Often conflicting reasons affecting a practical resolution may be comparable, and one of them overrides or defeats the others. But on other occasions we can regard reasons as guaranteed by different values which are commensurable though not through a precisely definable ranking like in decision theory, but following other patterns probably based on experience. So not all conflicts of reasons can be resolved by simply calculating the relative strength or force of each of them in order to determine which one defeats the other. Concerning this last issue Sloman (1990, p. 235) remarks that: "Some comparators apply constraint goals in planning, for instance using a "minimize cost" rule to select the cheaper of two sub-goals. Others directly order ends, like a rule that saving life is always more important than any other goal, but not because of some common measure applicable to both. As there are different incommensurable sources of motivation and different bases of comparison, there need not be any optimal resolution of a conflict". At this level then what is needed is a way of making comparisons among alternative potential goals that permits us to pick out only those goals that it would be reasonable for the agent to adopt according to his conative dispositions and doxastic states (cf. Pérez Miranda, 1997). And, clearly, to develop such a project we need something other than the original mathematical apparatus offered by rational choice theory.

---

produce goals for the benefit of others, and these can conflict with the individual's own goals and needs. Goal comparators are therefore needed for selection between different ends" (Sloman (1990, p. 234)).

2. GAME THEORY AND BEYOND

When the rational choice model of action is explicitly applied to social interactions it takes us into what is known as game theory. This theory analyses strategic rational choices in an ideal-type setting. Game theory is broadly conceived as an extension of rational choice theory for interactive or strategic decision-making situations. There are two ideas shared by these theories: the idea that rationality is utility maximization; and the idea that rational beliefs and rational expectations can be formalised using probability theory. There is, however, a third point that seems to distinguish one from the other: the idea that some strategy is necessary for situations of interaction in non-cooperative games. In its simplest form a *game* can be described as a situation in which the actions of one agent partially influence the welfare of another and *vice versa*. In these situations, an agent must assume that the other players make rational decisions in the same way as he does, so that mutual expectations concerning the rational decisions of others partly influence the rationality of the decision. Strategic situations are thus characterized by interdependence of decisions. That is, "Before making up his mind, each agent has to anticipate what others are likely to do, which may require an estimate of what they anticipate that he will do. His decision enters as part-determinant of the constraints that shape his decision" (Elster, 1986, p. 7). We focus our attention only on non-cooperative game theory. The aim of this theory is to explain how cooperation between rational and fully informed agents may take place in the absence of an ability to make binding agreements.

2.1 Equilibrium Points and the Common Knowledge of Rationality

In game theory some strategy is necessary for situations of interaction among agents. Each agent may need a strategy which takes account of other agents' strategies. To be sure, the expectations about the other players' decisions constitute an important part of the theory insofar as they are based on mutual beliefs about the others' rational behaviour.

One of the concerns of game theory is to give advice about how rational players can logically eliminate strategies that are strategically inferior. Game theory takes as rational decisions only those that are part of the *equilibrium points* of interaction. Among equilibrium points the most important and central one is probably *Nash equilibrium.* Informally, Nash equilibria are sets of strategies used by the 1 to *n* players in a game such that, for each player *i,* given the strategies of the other players, *i* has no incentive to change his strategy. However, application of this theory has revealed weak points, which has led to the project of refining it.[5]

The Common Knowledge of Rationality (henceforth, CKR) is an ideal game theory. This theory holds that agents are able to order their preferences in a consistent and complete way, and represent them by a ranking of the possible outcomes resulting from the situations of interaction. Players' information includes mutual belief (or common knowledge) that other players are also rational agents, so that if anything is known by an agent, this agent must know that the other agent knows what he knows, and *vice versa.* For instance, if there was any dominant strategy in the game, everyone would be able to draw it after a reasoning process. However, as we shall see the use of CKR raises difficulties when applied to several games. Even worse, in most games the recommendation "If there is a unique equilibrium for the game choose the equilibrium strategy" cannot be derived from the concept of optimising practical rationality (cf. Nida-Rümelin, 1994). Agents' beliefs can be internally consistent but fail to achieve the interpersonal consistency that makes it possible for one equilibrium to be singled out. In other words, the assumption of CKR concerning specific structures of interaction is in conflict with the theory of equilibrium.

2.2 Two Non-cooperative Games: Coordination and the Prisoner's Dilemma

The scope of the CKR is controversial in situations of interaction where agents, even though ideal rational players, make misleading decisions or are led to situations of infinite regress. A good way of looking at this issue is to check out paradigmatic "coordination" and "conflict" games.

Coordination: One feature of *coordination games* is that the utilities of coordinated strategies have higher values than those of uncoordinated strategies. The key point about coordination games (see below the *normal form* of the game Heads and Tails) is that the players have a mutual interest in coordinating. Since there is no conflict of interests, it would be odd if rational agents failed to find mutually beneficial solutions, at least with repeated play (cf. Hollis, 1994). Nevertheless, the problem is that according to the assumption of the CKR, to be rational, A's election should necessarily take B's election into account, but, to be rational B's election depends similarly on A's election. Therefore, the CKR leads agents to the problem of

[5] Our aim in this section is not so much to survey the degree of suitability of game theory concerning the elimination of alternatives that are strategically inferior or implausible, but to examine the extent to which this theory is suitable for the explanation of phenomena like coordination, cooperation, or the emergence of norms.

infinite regress. In short, the CKR fails to give good advice in decision-making processes in which agents need to coordinate their behaviours.[6]

Unfortunately the theory of equilibrium does not always provide a solution to the problem of coordination. In Heads and Tails there are two Nash equilibria in pure strategies[7] (and a third one in mixed strategies) and the players must each decide which to aim for. In this kind of coordination games with an equilibrium that is superior to the rest in the matrix, there is a unique solution that should apparently be selected by every agent. Indeed, in our coordination game the outcome (Heads, Heads) is better for both players, but it seems that agents cannot reach that conclusion without assuming something external to the CKR and to the theory of equilibrium.

		Player B	
		Heads	**Tails**
Player A	**Heads**	**2,2**	**0,0**
	Tails	**0,0**	**1,1**

What matters here is to explain the *method* agents use for choosing one of the coordinated solutions. In situations like the previous one where there is more than one equilibrium point, there are at least two ways in which one of them can be picked out as a solution. The first way of solving their choice problem is to use the *Principle of Coordination*: If one outcome is strictly preferred to every other outcome by both agents, then rationality requires each player to pick out the strategy which leads to that outcome. This principle, however, is not an implication of the CKR. If the principle of coordination is adhered to, each agent should expect that the other agent is following it as well. But such an expectation could only be justified to the extent that each agent thinks of the other as a utility maximizer (cf. Hollis & Sugden, 1993). The second way around the choice problem is to select one of the equilibria as the solution provided it has a certain perceptual salience in the minds of both players. That is, the equilibrium must be a "focal point" (Schelling, 1963). In more realistic situations, away from the abstractions of a pay-off matrix, players may be able to choose between equilibria using information that is not specified in the rules of the game. For instance, where these equilibria are symmetric one could by way of explanation claim that agents' decisions are grounded on *conventions* —that is, behavioural

[6] For some authors the CKR is nothing but an incoherent theory (cf. Hollis & Sugden, 1993).

[7] A pure strategy is one without elements of randomness; a mixed strategy, by contrast, is one in which a player randomizes between two or more strategies using probabilities.

conformities without explicit agreements among agents— but this resource clearly lies outside the nature of the CKR. The focal point effect defines both an essential limit on the ability of mathematical game theory to predict agents' behaviour in real cooperation or conflict situations and an important line for research in social domains. "It should not be surprising that game theory cannot provide a complete theory of human behavior without complementary theories from other disciplines. (….) The focal point effect of game theory can offer a useful perspective on the role of culture and perception. In particular, game theory can predict that cultural norms may be less important in situations where the set of equilibria is small (Myerson, 1991, p. 114).

Another possible line of analysis is to conceive the agents as a single unit of agency. In this case we are no longer considering strategic decision-making for a particular agent, but cooperative choice for a collective or group. Collective agency, however, requires deep revisions of the rational choice theory at the level of propositional attitudes. In particular, this idea of supra-individual agency implies the use of collective intentions, collective beliefs, and so on.[8]

Prisoner's dilemma: Where agents' interests are in conflict, it is not always easy to find a solution that satisfies everybody. The logical structure of an interactive game such as the prisoner's dilemma reveals problems concerning *equilibrium conflict*. This dilemma has drawn so much attention because it bears directly on fundamental issues in ethics and political philosophy and threatens the foundations of the social sciences.[9] The puzzle that emerges from this dilemma illustrates a conflict between individual and collective rationality. A collective whose components pursue rational self-interest may all end up worse off than a collective whose members act contrary to rational self-interest. The striking aspect of the (defect, defect) outcome in the prisoner's dilemma (see below) is that although it is the one we expect to arise, it is not the best outcome for the players jointly; both players would prefer that neither of them defect. For this reason, this dilemma is the paradigmatic example of self-interested, rational behaviour not leading to a socially optimal result.

The game theoretic decision to go for the *dominant strategy* in the prisoner's dilemma —i.e. to make a non-cooperative decision— is likely to be individually rational, but collectively irrational, since the outcome which is reached if both parties interact rationally from an individual point of view is worse than the outcome which would be reached if both adopted the other

[8] For theoretical work on collective agency consult Holmström-Hintikka & Tuomela (eds.) (1997).

[9] On this point see Campbell & Sowden (1985).

alternative. Agents' decisions based on dominant strategies are not invariably Pareto optimal. Or as L.H. Davis has put it: "But rationality has never been supposed to be guarantee of the best possible outcome; at least it would be a mistake to suppose that it is" (Davis, 1985, p. 47). The problem for the game-theoretic agent is that there is no *a priori* best strategy for him, since the success of his strategy depends on the strategy of the other player(s) and vice versa.

		Column Player	
		Cooperate	**Defect**
File Player	**Cooperate**	**2,2**	**0,0**
	Defect	**0,0**	**1,1**

In this game we expect an equilibrium where the combination of pay-offs is (1,1); and there is no need in arriving at this conclusion for either player to get entangled in thoughts about CKR. Each knows the best thing to do is "defect" and yet it yields a paradoxical result of making each worse off than they might have been had they each chosen "cooperate" and so obtained one result in which the pay-offs are (2,2).

Note that communication between agents doesn't resolve the conflict that emerges from the prisoner's dilemma. If they could communicate with one another, they would quickly come to agree that the best for both comes from "cooperating". However, is it in the interest of each party to keep to such an agreement? No, the best answer in terms of pay-offs is still to "defect". What seems to be required to avoid this outcome is a *mechanism for enforcing an agreement*, thus assuring that both actually do "cooperate". However the game is non-cooperative in nature and, consequently, players cannot make enforceable agreements to cooperate. Moreover, a promise to cooperate must be credible in utilitarian terms. What would happen if the game was repeated several times? Both players know that in the last *N*th round both will defect, no matter what they did before. So they might as well defect in the *N*–1th round, too. Proceeding inductively, both players defect in each period. If by contrast the game is infinitely iterated (or the players do not know which round is the last) the inductive argument that leads agents to defect at each stage does not hold. In such games mutual cooperation is possible (Axelrod, 1984). Cooperation can also be achieved in finitely repeated games of incomplete information where players are uncertain about other players" rationality (see, Kreps *et al.*, 1982). This last result partially confirms experimental evidence about cooperative behaviour in situations where agents known that the game has a finite number of rounds.

2.3 Norms and Procedural Rationality

Based upon methodological individualism rational choice theorists and game theorists have attempted to account for all sort of macro-phenomena, such as political action or norms, as emergent properties of rational individual choices. Indeed, some problems of coordination and cooperation among agents in game theory can be seen as a reason for the spreading of some type of norms, but the game-theoretic analysis doesn't account for all manner of social norms; it must probably presuppose at least some of them. Though several spontaneous forms of social order may be explained from a game-theoretic framework, without the necessity of explicit agreement, other social individual behaviours are "guided" by prescriptions that clearly go beyond the scope of game theory. People follow norms not only because of the benefits obtained from doing so, but also to prevent the cost of possible penalties and for many other reasons as well. For instance, *norms of cooperation* are closely related to the n-person prisoner's dilemma, where each player has a dominant strategy that he should follow, according to the normative principles of rational choice theory and independently of other players' elections. Since the game is non-cooperative there is no external authority to control and enforce sanctions to defectors. "It would be better for everybody to cooperate, but since cooperation is a dominated strategy, any agreement to cooperate would fail to be self-enforcing, since each player would have an incentive to cheat on the others" (Bicchieri 1997, p. 19). There are nevertheless occasions on which people tend to cooperate disregarding the dominant strategy. For instance, if one lives in a small village where one is not anonymous, it will usually be preferable to cooperate and preserve one's reputation, whereas one might act otherwise in other circumstances. Therefore, to explain why cooperation or defection might be expected into a collective, we need to look at the circumstances underlying agents' decisions and not only to the dictates of the rational choice theory.

Agents' decisions are also influenced by *symbolic utilities* that are not necessarily embedded in situations of equilibrium conflict (cf. Nozick, 1993). Agents follow different kinds of norms, rules, and routines to which they generally get familiar with through learning processes. The shared rules of procedural rationality provide an irreducible social and historical context within which instrumentally rational actions can be undertaken. Since all our process of practical rationality generally rest on custom, it is no surprise if regularities in behaviour are to be explained by appealing to custom and not only to utilitarian deliberation. Procedural rationality marks the limits of instrumental rationality (cf. Hargreaves *et al.* 1992). In addition to it, norms may generate reasons for action which cannot be assimilated to some prior set of objectives. When specific conditions hold, many of these norms are automatically triggered without the influence of deliberating mechanisms.

At the same time we should bear in mind that the transgression and modification of norms is also an important part of our everyday social behaviour. Thus, it is the agent who interprets and, occasionally, violates established norms. As we have already pointed out the existing rules and norms for coordinating social behaviours provide a context for the decisions made by the agent. It can be rational to do something that fits in an approved structure of individual or social activity. However, individual agents as well as groups are able to modify and, sometimes, transgress those norms to suit their own preferences. This seems to indicate, first, that social structures that condition agents' behaviour are dynamic enough to be modified, if it is required; and, second, that structural rationality might be somehow compatible with an act-utilitarian version of the rational choice paradigm. Nonetheless, this last issue is more controversial than it could appear at first glance. Consider, for instance, the practice of promising. Act-utilitarianism is often alleged to recognise no binding force to promises at all, and to hold that promises should be kept only when doing the thing promised is likely to produce more utility than any alternative actions. But promising seems to be a useful practice only if agents desist from thinking in utilitarian terms and agree to abide by the rules which are constitutive of the practice. To be sure, a paradigm of "structural" rationality can very easily come into conflict with the most basic requirements of rational choice theory.

Some social scientists have pointed out that in game theory the agents' reasons for actively engaging in social actions are ignored. "Therefore, while macro-social phenomena are allowed to emerge from the agents' choices, the mental mechanisms of micro-interactions are taken for granted. They always pre-exist and explain the macro-social level, but are never shown to be produced or influenced by it" (Conte & Castelfranchi, 1995, p. 6). In addition, it has been observed that the game-theoretic structure does not account for social action aimed at modifying the behaviour and mental states of other agents, although this type of social action is no less important than the game-theoretic one (cf. Ibid.). If we agree with the view that society consists of individuals plus the structural and objective relations that determine the behaviour of its members, we should look at the old problem of the micro-macro gap through a different lens. Specifically, not only should the way down of the micro-macro links be accounted for, but also the way up, i.e. not only should the direction from the individual to the group be studied, but also from the group to the individual.

2.4 Collective Decision-making

It is convenient to distinguish between interactive decision-making and collective decision-making. In a collective choice the decision of some agents exerts an influence on all. That is, collective choice is characterized by the fact that it holds for all members of a group. A great amount of work has been done on the problem of how to put together preferences of individuals in order to obtain a reasonable collective choice. However, results derived from several works on the nature of the logic of collectives have undermined the idea that social action is a sum of individually optimising behaviour (in the sense of the original rational choice-paradigm), and that collective action can be ascribed to an optimising collective agent (cf. Arrow 1951; Sen 1970). From these works we draw the conclusion that the utility function of an aggregate is not reducible to the sum of the utility functions of all its members. In the case we accepted that reduction, we would have to face Arrow's Impossibility Theorem. This theorem entails that a collective utility function based on the rational choice principle leads us to conflict and paradox situations. In other words, no social-welfare function is possible which satisfies a number of prima-facie highly reasonable assumptions about such a function. Thus, there can be no ideally rational aggregation device (see e.g. Arrow 1950; Vickrey 1960; Mackay 1980). "If Arrow is correct, then we cannot safely trust to our intuitions to show us how to construct a satisfactory social-welfare function" (Scarre 1996, p. 19). In any case, if collective choice is not a preference aggregation process, what kind of process would it be? Could it be a dialogical process where the reasons of each agent in the group are exchanged in the way of looking for a consensus? An adequate answer to such questions will probably only be possible with a much more modest conception of rationality in theories of agent interaction.

FINAL REMARKS

First, as we have already seen for cases of strategic situation the goal is not simply to choose rationally, but to choose in such a way that a mutual solution is achieved. Current solution concepts, especially those concerning strategic interactions and rational expectations, are based on an implicit assumption that individuals know the prevailing equilibrium. It is one thing to know that an equilibrium exists and another to have reasons to think that it will be realised by limited agents (see Elster, 1986). Moreover, leaving aside the problems related to limited agents, decision-makers also have to take into account the social setting in which they perform their actions. Structural rationality is not only useful in resolving equilibrium selection problems in games, it is also essential in the explanation of agents' behaviour. We should remember that many of our decisions are clearly influenced by social and cultural norms far beyond any game-theoretical analysis. That is why together with the reformulation and refinement of the notion of equilibrium, the role of society and culture can no longer be marginalised in these studies. Indeed, the existing gap between instrumental and structural rationality might be resolved if we manage to strike a balance between the dispositional conditioning which restricts the number of possible actions of the agent, on the one hand, and the decision-making mechanisms, on the other. It is worth noting that both individual and collective actions are fitted into a socially structured network where the set of possible strategies permitted is to a certain extent generally given. In brief, the rational choice paradigm seems to require a deep revision in the direction of structural rationality.

Second, if we are interested in answering questions like what are the kinds of reasoning and decision-making procedures that people actually use and why, then it would be better to adopt a different framework and methodology to explain behaviour under reasonable conditions. In our view cognition must be seen as taking part not only in the brain, but in interaction with the body and the surrounding world. Both empirical (or experimental) observations and artificial systems give us a good testbed for our theories on decision-making (cf. Pérez Miranda, 2001). There is no doubt that ideal and "armchair" models of reasoning and decision-making are being replaced by less idealised models sensitive to experimental data, or by models that find their scientific support in accurate computer simulation.

ACKNOWLEDGMENTS

This paper has been supported by the Research Project 1/UPV/EHU 00I09-HA-4487/1998.

REFERENCES

Allais, M. (1953), 'Le comportement de l'homme rationnel devant le risque: critique des postulats et axiomes de l'école améficaine'. *Econometrica* 21: 503-46.

Anand, P. (1993), *Foundations of Rational Choice Under Risk.* Oxford, Oxford University Press.

Arrow, K.J. (1950), 'A difficulty in the concept of social welfare'. *Journal of Political Economy*, 58: 328-46.

Arrow, K. J. (1951), *Social Choice and Individual Values.* New York, Wiley & Sons.

Axelrod, R. (1984), *The evolution of cooperation.* New York, Basic Books.

Bicchieri, C. (1997), "Learning to cooperate". In *The Dynamics of Norms*, C. Bicchieri *et al.* (eds.). Cambridge, Cambridge University Press.

Campbell, R. & L. Sowden (eds.) (1985), *Paradoxes of Rationality and Cooperation*, Vancouver, The University of British Columbia Press.

Cherniak, C. (1986), *Minimal Rationality*. Cambridge, MA, The MIT Press.

Collins, R. (1994), *Four Sociological Traditions*. Oxford, Oxford University Press.

Conte, R. & Castelfranchi, C. (1995), *Cognitive and Social Action*. London, UCL, Press.

Davis, L.H. (1985), "Prisoners, Paradox, and Rationality". In *Paradoxes of Rationality and Cooperation*, R. Campbell & L. Sowden (eds.) (1985). Vancouver, The University of British Columbia Press.

DiFinetti, B. (1975), *Theory of Probability*. New York, Wiley.

Doyle, J. (1992), 'Rationality and its Role in Reasoning'. *Computational Intelligence*, 8 (2), 376-409.

Ellsberg, D. (1961), 'Risk, Ambiguity, and the Savage Axioms'. *Quartely Journal of Economics* 75: 643-69.

Elster, J. (ed.) (1986), *Rational Choice. Readings in Social and Political Theory*. Oxford, Basil Blackwell.

Hargreaves, S. (1998), "Expressive rationality: is self worth just another kind of preference?" To appear in *Ontology in Economics*, U. Maki (ed.). Cambridge, University Press.

Hargreaves, S. *et al.* (1992), *The Theory of Choice*. Oxford, Blackwell.

Hollis, M. & S. Sudgen (1993), 'Rationality in Action'. *Mind* 405, 1-35.

Hollis, M. (1994), *The Philosophy of Social Sciences*. Cambridge, Cambridge University Press.

Holmström-Hintikka G. & R. Tuomela (eds.) (1997), *Contemporary Action Theory. Volume 2: Social Action*. Dordrecht, Kluwer Academic Publishers.

Jeffrey, R. (1983), *The Logic of Decision*. Chicago, Chicago University Press.

Kreps, D. *et al.* (1982), 'Rational Cooperation in the Finitely Repeated Prisoner's Dilemma'. *Journal of Economic Theory* 27: 245-52.

Kyburg, H. (1991), "Normative and Descriptive Ideals". In *Philosophy and AI. Essays at the Interface*, R. Cummins & J. Pollock (eds.) (1991). Cambridge, MA, The MIT Press.

Mackay, A.F. (1980), *Arrow's Theorem: The Problem of the Social Choice*. New Haven and London, Yale University Press.

Myerson, R.B. (1991), *Game Theory*. Harvard, Harvard University Press.

Nida-Rümellin, J. (1994), 'Rational Choice: Extensions and Revisions'. *Ratio* (New Series) VII 2: 122-44.

Nozick, R. (1993), *The Nature of Rationality*. Princeton, New Jersey, Princeton University Press.

Pérez Miranda, L.A. (1997), 'Deciding, Planning, and Practical Reasoning'. *Argumentation* 11 (4): 435-461.

Pérez Miranda, L.A. (2001), 'Rational approaches to rationality'. *Trends in Cognitive Sciences* V (8): 372-73.

Rubinstein, A. (1998), *Modeling Bounded Rationality*. Cambridge, MA, The MIT Press.

Savage, J. L. (1972), *The Foundations of Statistics*. New York, Elsevier.

Schelling, T.C. (1963), *The Strategy of Conflict*. Cambridge, MA, Harvard University Press.

Scarre, G. (1996), *Utilitarianism*. London, Routledge.

Sen, A.K. (1970), *Collective Choice and Social Welfare*. San Francisco, Holden-Day.

Skyrms, B. (1980), *Causal Necessity*. New Haven, Yale University Press.

Sloman, A. (1990), "Motives, mechanisms, and emotions". In *The Philosophy of Artificial Intelligence,* M.A. Boden (ed.). Oxford Readings in Philosophy Series, Oxford University Press, 231-247.

Tversky, A. & Kahneman, D. (1974), 'Judgment under uncertainty: heuristics and biases'. *Science* 185, 1124-31.

Vickrey, W. (1960), 'Utility, Strategy, and Social Decisions Rules'. *The Quarterly Journal of Economics*, 4: 507-535.

von Neumann, J. & Morgenstern, O. (1944), *Theory of Games and Economic Behavior.* New York, Wiley.

Acknowledgments

ICCS-01 was financially supported by the Secretary of State of Science and Technology of the Spanish Government, the Secretary of Education, Universities, and Research of the Basque Government, the Kutxa Foundation, the Episteme Foundation, and Iberdrola Electric Company. We would like to thank them all. We would also like to thank the members of the organizing and program committees for their invaluable work before and during the colloquium. As for the preparation of this volume, we are grateful to an anonymous referee and the responsibles from Kluwer, Keith Lehrer, Rudolf Rijgersberg, and Ingrid Krabbenbos. And last, but not least, thanks also to Ixiar Bidaola from the ILCLI for her help in typing and editing some papers.

Index

PHILOSOPHICAL STUDIES SERIES

1. Jay F. Rosenberg: *Linguistic Representation.* 1974 ISBN 90-277-0533-X
2. Wilfrid Sellars: *Essays in Philosophy and Its History.* 1974 ISBN 90-277-0526-7
3. Dickinson S. Miller: *Philosophical Analysis and Human Welfare.* Selected Essays and Chapters from Six Decades. Edited with an Introduction by Lloyd D. Easton. 1975 ISBN 90-277-0566-6
4. Keith Lehrer (ed.): *Analysis and Metaphysics.* Essays in Honor of R. M Chisholm. 1975 ISBN 90-277-0571-2
5. Carl Ginet: *Knowledge, Perception, and Memory.* 1975 ISBN 90-277-0574-7
6. Peter H. Hare and Edward H. Madden: *Causing, Perceiving and Believing.* An Examination of the Philosophy of C. J. Ducasse. 1975 ISBN 90-277-0563-1
7. Hector-Neri Castañeda: *Thinking and Doing.* The Philosophical Foundations of Institutions. 1975 ISBN 90-277-0610-7
8. John L. Pollock: *Subjunctive Reasoning.* 1976 ISBN 90-277-0701-4
9. Bruce Aune: *Reason and Action.* 1977 ISBN 90-277-0805-3
10. George Schlesinger: *Religion and Scientific Method.* 1977 ISBN 90-277-0815-0
11. Yirmiahu Yovel (ed.): *Philosophy of History and Action.* Papers presented at the First Jerusalem Philosophical Encounter (December 1974). 1978 ISBN 90-277-0890-8
12. Joseph C. Pitt (ed.): *The Philosophy of Wilfrid Sellars: Queries and Extensions.* 1978 ISBN 90-277-0903-3
13. Alvin I. Goldman and Jaegwon Kim (eds.): *Values and Morals.* Essays in Honor of William Frankena, Charles Stevenson, and Richard Brandt. 1978 ISBN 90-277-0914-9
14. Michael J. Loux: *Substance and Attribute.* A Study in Ontology. 1978 ISBN 90-277-0926-2
15. Ernest Sosa (ed.): *The Philosophy of Nicholas Rescher.* Discussion and Replies. 1979 ISBN 90-277-0962-9
16. Jeffrie G. Murphy: *Retribution, Justice, and Therapy.* Essays in the Philosophy of Law. 1979 ISBN 90-277-0998-X
17. George S. Pappas (ed.): *Justification and Knowledge.* New Studies in Epistemology. 1979 ISBN 90-277-1023-6
18. James W. Cornman: *Skepticism, Justification, and Explanation.* With a Bibliographic Essay by Walter N. Gregory. 1980 ISBN 90-277-1041-4
19. Peter van Inwagen (ed.): *Time and Cause.* Essays presented to Richard Taylor. 1980 ISBN 90-277-1048-1
20. Donald Nute: *Topics in Conditional Logic.* 1980 ISBN 90-277-1049-X
21. Risto Hilpinen (ed.): *Rationality in Science.* Studies in the Foundations of Science and Ethics. 1980 ISBN 90-277-1112-7
22. Georges Dicker: *Perceptual Knowledge.* An Analytical and Historical Study. 1980 ISBN 90-277-1130-5
23. Jay F. Rosenberg: *One World and Our Knowledge of It.* The Problematic of Realism in Post-Kantian Perspective. 1980 ISBN 90-277-1136-4
24. Keith Lehrer and Carl Wagner: *Rational Consensus in Science and Society.* A Philosophical and Mathematical Study. 1981 ISBN 90-277-1306-5
25. David O'Connor: *The Metaphysics of G. E. Moore.* 1982 ISBN 90-277-1352-9

PHILOSOPHICAL STUDIES SERIES

26. John D. Hodson: *The Ethics of Legal Coercion.* 1983 ISBN 90-277-1494-0
27. Robert J. Richman: *God, Free Will, and Morality.* Prolegomena to a Theory of Practical Reasoning. 1983 ISBN 90-277-1548-3
28. Terence Penelhum: *God and Skepticism.* A Study in Skepticism and Fideism. 1983 ISBN 90-277-1550-5
29. James Bogen and James E. McGuire (eds.): *How Things Are.* Studies in Predication and the History of Philosophy of Science. 1985 ISBN 90-277-1583-1
30. Clement Dore: *Theism.* 1984 ISBN 90-277-1683-8
31. Thomas L. Carson: *The Status of Morality.* 1984 ISBN 90-277-1619-9
32. Michael J. White: *Agency and Integrality.* Philosophical Themes in the Ancient Discussions of Determinism and Responsibility. 1985 ISBN 90-277-1968-3
33. Donald F. Gustafson: *Intention and Agency.* 1986 ISBN 90-277-2009-6
34. Paul K. Moser: *Empirical Justification.* 1985 ISBN 90-277-2041-X
35. Fred Feldman: *Doing the Best We Can.* An Essay in Informal Deontic Logic. 1986 ISBN 90-277-2164-5
36. G. W. Fitch: *Naming and Believing.* 1987 ISBN 90-277-2349-4
37. Terry Penner: *The Ascent from Nominalism.* Some Existence Arguments in Plato's Middle Dialogues. 1987 ISBN 90-277-2427-X
38. Robert G. Meyers: *The Likelihood of Knowledge.* 1988 ISBN 90-277-2671-X
39. David F. Austin (ed.): *Philosophical Analysis.* A Defense by Example. 1988 ISBN 90-277-2674-4
40. Stuart Silvers (ed.): *Rerepresentation.* Essays in the Philosophy of Mental Representation. 1988 ISBN 0-7923-0045-9
41. Michael P. Levine: *Hume and the Problem of Miracles.* A Solution. 1989 ISBN 0-7923-0043-2
42. Melvin Dalgarno and Eric Matthews (eds.): *The Philosophy of Thomas Reid.* 1989 ISBN 0-7923-0190-0
43. Kenneth R. Westphal: *Hegel's Epistemological Realism.* A Study of the Aim and Method of Hegel's *Phenomenology of Spirit.* 1989 ISBN 0-7923-0193-5
44. John W. Bender (ed.): *The Current State of the Coherence Theory.* Critical Essays on the Epistemic Theories of Keith Lehrer and Laurence BonJour, with Replies. 1989 ISBN 0-7923-0220-6
45. Roger D. Gallie: *Thomas Reid and 'The Way of Ideas'.* 1989 ISBN 0-7923-0390-3
46. J-C. Smith (ed.): *Historical Foundations of Cognitive Science.* 1990 ISBN 0-7923-0451-9
47. John Heil (ed.): *Cause, Mind, and Reality.* Essays Honoring C. B. Martin. 1989 ISBN 0-7923-0462-4
48. Michael D. Roth and Glenn Ross (eds.): *Doubting.* Contemporary Perspectives on Skepticism. 1990 ISBN 0-7923-0576-0
49. Rod Bertolet: *What is Said.* A Theory of Indirect Speech Reports. 1990 ISBN 0-7923-0792-5
50. Bruce Russell (ed.): *Freedom, Rights and Pornography.* A Collection of Papers by Fred R. Berger. 1991 ISBN 0-7923-1034-9
51. Kevin Mulligan (ed.): *Language, Truth and Ontology.* 1992 ISBN 0-7923-1509-X

PHILOSOPHICAL STUDIES SERIES

52. Jesús Ezquerro and Jesús M. Larrazabal (eds.): *Cognition, Semantics and Philosophy.* Proceedings of the First International Colloquium on Cognitive Science. 1992 ISBN 0-7923-1538-3
53. O.H. Green: *The Emotions.* A Philosophical Theory. 1992 ISBN 0-7923-1549-9
54. Jeffrie G. Murphy: *Retribution Reconsidered.* More Essays in the Philosophy of Law. 1992 ISBN 0-7923-1815-3
55. Phillip Montague: *In the Interests of Others.* An Essay in Moral Philosophy. 1992 ISBN 0-7923-1856-0
56. Jacques-Paul Dubucs (ed.): *Philosophy of Probability.* 1993 ISBN 0-7923-2385-8
57. Gary S. Rosenkrantz: *Haecceity.* An Ontological Essay. 1993 ISBN 0-7923-2438-2
58. Charles Landesman: *The Eye and the Mind.* Reflections on Perception and the Problem of Knowledge. 1994 ISBN 0-7923-2586-9
59. Paul Weingartner (ed.): *Scientific and Religious Belief.* 1994 ISBN 0-7923-2595-8
60. Michaelis Michael and John O'Leary-Hawthorne (eds.): *Philosophy in Mind.* The Place of Philosophy in the Study of Mind. 1994 ISBN 0-7923-3143-5
61. William H. Shaw: *Moore on Right and Wrong.* The Normative Ethics of G.E. Moore. 1995 ISBN 0-7923-3223-7
62. T.A. Blackson: *Inquiry, Forms, and Substances.* A Study in Plato's Metaphysics and Epistemology. 1995 ISBN 0-7923-3275-X
63. Debra Nails: *Agora, Academy, and the Conduct of Philosophy.* 1995 ISBN 0-7923-3543-0
64. Warren Shibles: *Emotion in Aesthetics.* 1995 ISBN 0-7923-3618-6
65. John Biro and Petr Kotatko (eds.): *Frege: Sense and Reference One Hundred Years Later.* 1995 ISBN 0-7923-3795-6
66. Mary Gore Forrester: *Persons, Animals, and Fetuses.* An Essay in Practical Ethics. 1996 ISBN 0-7923-3918-5
67. K. Lehrer, B.J. Lum, B.A. Slichta and N.D. Smith (eds.): *Knowledge, Teaching and Wisdom.* 1996 ISBN 0-7923-3980-0
68. Herbert Granger: *Aristotle's Idea of the Soul.* 1996 ISBN 0-7923-4033-7
69. Andy Clark, Jesús Ezquerro and Jesús M. Larrazabal (eds.): *Philosophy and Cognitive Science: Categories, Consciousness, and Reasoning.* Proceedings of the Second International Colloquium on Cogitive Science. 1996 ISBN 0-7923-4068-X
70. J. Mendola: *Human Thought.* 1997 ISBN 0-7923-4401-4
71. J. Wright: *Realism and Explanatory Priority.* 1997 ISBN 0-7923-4484-7
72. X. Arrazola, K. Korta and F.J. Pelletier (eds.): *Discourse, Interaction and Communication.* Proceedings of the Fourth International Colloquium on Cognitive Science. 1998 ISBN 0-7923-4952-0
73. E. Morscher, O. Neumaier and P. Simons (eds.): *Applied Ethics in a Troubled World.* 1998 ISBN 0-7923-4965-2
74. R.O. Savage: *Real Alternatives, Leibniz's Metaphysics of Choice.* 1998 ISBN 0-7923-5057-X
75. Q. Gibson: *The Existence Principle.* 1998 ISBN 0-7923-5188-6
76. F. Orilia and W.J. Rapaport (eds.): *Thought, Language, and Ontology.* 1998 ISBN 0-7923-5197-5

PHILOSOPHICAL STUDIES SERIES

77. J. Bransen and S.E. Cuypers (eds.): *Human Action, Deliberation and Causation.* 1998 ISBN 0-7923-5204-1
78. R.D. Gallie: *Thomas Reid: Ethics, Aesthetics and the Anatomy of the Self.* 1998 ISBN 0-7923-5241-6
79. K. Korta, E. Sosa and X. Arrazola (eds.): *Cognition, Agency and Rationality.* Proceedings of the Fifth International Colloquium on Cognitive Science. 1999 ISBN 0-7923-5973-9
80. M. Paul: *Success in Referential Communication.* 1999 ISBN 0-7923-5974-7
81. E. Fischer: *Linguistic Creativity.* Exercises in 'Philosophical Therapy'. 2000 ISBN 0-7923-6124-5
82. R. Tuomela: *Cooperation.* A Philosophical Study. 2000 ISBN 0-7923-6201-2
83. P. Engel (ed.): *Believing and Accepting.* 2000 ISBN 0-7923-6238-1
84. W.L. Craig: *Time and the Metaphysics of Relativity.* 2000 ISBN 0-7923-6668-9
85. D.A. Habibi: *John Stuart Mill and the Ethic of Human Growth.* 2001 ISBN 0-7923-6854-1
86. M. Slors: *The Diachronic Mind.* An Essay on Personal Identity, Psychological Continuity and the Mind-Body Problem. 2001 ISBN 0-7923-6978-5
87. L.N. Oaklander (ed.): *The Importance of Time.* Proceedings of the Philosophy of Time Society, 1995–2000. 2001 ISBN 1-4020-0062-6
88. M. Watkins: *Rediscovering Colors.* A Study in Pollyanna Realism. 2002 ISBN 1-4020-0737-X
89. W.F. Vallicella: *A Paradigm Theory of Existence.* Onto–Theology Vindicated. 2002 ISBN 1-4020-0887-2
90. M. Hulswit: *From Cause to Causation.* A Peircean Perspective. 2002 ISBN 1-4020-0976-3; Pb 1-4020-0977-1
91. D. Jacquette (ed.): *Philosophy, Psychology, and Psychologism.* Critical and Historical Readings on the Psychological Turn in Philosophy. 2003 ISBN 1-4020-1337-X
92. G. Preyer, G. Peter and M. Ulkan (eds.): *Concepts of Meaning.* Framing an Integrated Theory of Linguistic Behavior. 2003 ISBN 1-4020-1329-9
93. W. de Muijnck: *Dependencies, Connections, and Other Relations.* A Theory of Mental Causation. 2003 ISBN 1-4020-1391-4
94. N. Milkov: *A Hundred Years of English Philosophy.* 2003 ISBN 1-4020-1432-5
95. E.J. Olsson (ed.): *The Epistomology of Keith Lehrer.* 2003 ISBN 1-4020-1605-0
96. D.S. Clarke: *Sign Levels.* Language and Its Evolutionary Antecedents. 2003 ISBN 1-4020-1650-6
97. A. Meirav: *Wholes, Sums and Unities.* 2003 ISBN 1-4020-1660-3
98. C.H. Conn: *Locke on Esence and Identity.* 2003 ISBN 1-4020-1670-0
99. J.M. Larrazabal and L.A. Pérez Miranda: *Language, Knowledge, and Representation. Proceedings of the Sixth International Colloquium on Cognitive Science (LCC-99).* 2004 ISBN 1-4020-2057-0
100. P. Ziff: *Moralities.* A Diachronic Evolution Approach. 2004 ISBN 1-4020-1891-6
101. J.A. Corlett: Terorism: *A Philosophical Analysis.* 2003 ISBN 1-4020-1694-8; Pb 1-4020-1695-6

PHILOSOPHICAL STUDIES SERIES

102. K. Korta and J.M. Larrazabal (eds.): *Truth, Rationality, Cognition, and Music*. Proceedings of the Seventh International Colloquium on Cognitive Science. 2003 ISBN 1-4020-1912-2

KLUWER ACADEMIC PUBLISHERS – DORDRECHT / BOSTON / LONDON